Facility Pollution Prevention Guide

Facility Pollution Prevention Guide

United States Environmental Protection Agency (EPA)

GLOBAL PROFESSIONAL PUBLICATIONS

Global Engineering Documents

IRVINE   WASHINGTON   ST. LOUIS   MIAMI   PARIS   HONG KONG

An IHS Group Company

Published by—

GLOBAL PROFESSIONAL PUBLICATIONS

Global Engineering Documents

2805 MCGAW AVENUE
IRVINE, CALIFORNIA 92714

FOREWORD

Today's rapidly changing technologies and industrial products and practices carry the risk of generating materials that, if improperly managed, can threaten public health and the environment. With the Pollution Prevention Act of 1990, the U.S. Congress established pollution prevention as a "national objective" and the most important component of the environmental management hierarchy. Thus, national policy declares that the creation of potential pollutants should be prevented or reduced during the production cycle whenever feasible.

In carrying out its program to encourage the adoption of Pollution Prevention, the Risk Reduction Engineering Laboratory and the Office of Solid Waste offer this *Facility Pollution Prevention Guide*. The *Guide's* predecessor, the *Waste Minimization Opportunity Assessment Manual*, published in 1988, concentrated primarily on the waste types covered in the Resource Conservation and Recovery Act (RCRA). In contrast, this edition deals with "multimedia" pollution prevention. This reflects our national realization, as demonstrated in the 1990 legislation, that we must look at wastes more broadly if we are to protect the environment adequately. That is, it is important to minimize all pollutants, including air emissions, wastewater discharges, and solid wastes as well as energy and water consumption. In addition to controlling waste creation during the production process, we need to design products that will have less impact on the environment while in use and after disposal.

This edition of the *Guide* is written for those individuals responsible for implementing pollution prevention in their facilities. It is intended to help small- to medium-sized production facilities develop broad-based, multimedia pollution prevention programs. It describes how to identify, assess, and implement opportunities for preventing pollution and how to stimulate the ongoing search for such opportunities. Companies that adopt this approach typically find that they reduce both their operating costs and their potential liabilities, in addition to helping to preserve the environment.

This is not intended to be a prescriptive, comprehensive document. It is necessarily a generalized approach, since it is intended for use by companies in all business and geographic areas. You are in the best position to judge how to develop a program that will fit your circumstances. We have addressed the basic steps involved in developing an adequate pollution prevention program. The true success of your efforts will be determined by the extent to which you are able to go beyond these basics. Because we *strongly encourage* you to go beyond a minimal program, this *Guide* also provides references and information sources that will help you expand your efforts.

ABSTRACT

The U.S. Environmental Protection Agency (U.S. EPA) developed the *Facility Pollution Prevention Guide* for those who are interested in and responsible for pollution prevention in industrial or service facilities. It summarizes the benefits of a company-wide pollution prevention program and suggests ways to incorporate pollution prevention in company policies and practices.

The *Guide* describes how to establish a company-wide pollution prevention program. It outlines procedures for conducting a preliminary assessment to identify opportunities for waste reduction or elimination. Then, it describes how to use the results of this preassessment to prioritize areas for detailed assessment, how to use the detailed assessment to develop pollution prevention options, and how to implement those options that withstand feasibility analysis.

Methods of evaluating, adjusting, and maintaining the program are described. Later chapters deal with cost analysis for pollution prevention projects and with the roles of product design and energy conservation in pollution prevention.

Appendices consist of materials that will support the pollution prevention effort: assessment worksheets, sources of additional information, examples of evaluative methods, and a glossary.

The draft information used for this Guide was compiled and prepared by Battelle, Columbus, Ohio, under Contract No. 68-CO-0003 for the U.S. EPA's Office of Research and Development.

CONTENTS

APPENDICES

ACKNOWLEDGEMENTS

This *Guide* was prepared under the direction and coordination of Lisa Brown of the U.S. Environmental Protection Agency (U.S. EPA), Pollution Prevention Research Branch, Risk Reduction Engineering Laboratory, Cincinnati, Ohio.

Battelle compiled and prepared the information used for this *Guide* under the direction of Bob Olfenbuttel. Participating in this effort for Battelle were Larry Smith, David Evers, Lynn Copley-Graves, Carol Young, and Sandra Clark.

Contributions were made by U.S. EPA's Office of Research and Development, the U.S. EPA Office of Solid Waste, the pollution prevention organizations in the U.S. EPA Regional Offices, state pollution prevention organizations, and members of academia and industry.

Specifically, the following people provided significant assistance:

Patrick Pesacreta
Office of Solid Waste
U.S. Environmental Protection Agency

Gary Hunt
North Carolina Office of Waste Reduction

Deborah Hanlon & Martin Spitzer
Pollution Prevention Division
U.S. Environmental Protection Agency

Abby Swaine
Region I Pollution Prevention Program
U.S. Environmental Protection Agency

Thomasine Bayless
Risk Reduction Engineering Laboratory
U.S. Environmental Protection Agency

Contributions to the development of this *Guide* were also made by the following people:

Alan Rimer
Alliance Technologies Corporation

Eugene B. Pepper
Office of Environmental Coordination
State of Rhode Island and Providence
 Plantations

Harry W. Edwards
Colorado State University

David L. Thomas, Ph.D.
Hazardous Waste Research and Information
 Center

Azita Yazdani
Pollution Prevention International

David M. Benforado
3M Corporation

R. Lee Byers
Aluminum Company of America

James R. Aldrich
University of Cincinnati

Henry W. Nowick
Envirocorp

James Edward
Pollution Prevention Division
U.S. Environmental Protection Agency

Chet McLaughlin
Region VII
U.S. Environmental Protection Agency

Marvin Fleischman & Clay Hansen
University of Louisville

Charles A. Pittinger, PhD
The Procter & Gamble Company

H. Lanier Hickman, Jr.
GRCDA/SWANA

Dent Williams
DIPEC

Charles Wentz
Argonne National Laboratory

Linda G. Pratt
San Diego County Department of Health
Services

Bruce Cranford
U.S. Department of Energy

L. M. Fischer
Allied-Signal

Thomas R. Hersey, Jr.
Erie County Pollution Prevention Program

Richard F. Nowina
Ontario Waste Management Corporation

David Hartley & Robert Ludwig
California Department of Toxic Substance
  Control

Bob Carter
Waste Reduction Resource
  Center — Southeast

Terry Foecke
WRITAR

Audun Amundsen
Stiftelsen Østfoldforskning
Norway

Birgitte B. Nielsen
Rendan A/S
Denmark

Michel Suijkerbuijk
Innovatiecentrum Overijssel
Netherlands

Per Kirkebak
Peterson A·S
Norway

Han Brezet & Bas Kothuis
TME
Netherlands

Sybren de Hoo
NOTA
Netherlands

Special acknowledgement is given to all
members of the Pollution Prevention Research
Branch, especially:

Ruth Corn, Rita Bender,
Harry Freeman, Ivars Licis,
Paul Randall, Mary Ann
Curran and Anne Robertson.

International contributions were made by:

Bärbel Hegenbart & Stefan Millonig
IÖW & VÖW
Austria

Brian Pearson
Aspects International Ltd.
England

Thomas Gutwinski
BAUM
Austria

Pollution prevention is the use of materials, processes, or practices that reduce or eliminate the creation of pollutants or wastes at the source. It includes practices that reduce the use of hazardous and nonhazardous materials, energy, water, or other resources as well as those that protect natural resources through conservation or more efficient use.

A pollution prevention program is an ongoing, comprehensive examination of the operations at a facility with the goal of minimizing all types of waste products. An effective pollution prevention program will:

- reduce risk of criminal and civil liability
- reduce operating costs
- improve employee morale and participation
- enhance company's image in the community
- protect public health and the environment.

This *Guide* is intended to assist you in developing a pollution prevention program for your business. It will help you decide which aspects of your operation you should assess and how detailed this assessment should be.

This chapter provides background information on pollution prevention. Specifically, it

- Summarizes the benefits you can obtain from a company-wide pollution prevention program that integrates raw materials, supplies, chemicals, energy, and water use.
- Describes the U.S. EPA's Environmental Management Hierarchy.
- Explains what pollution prevention is and what it is not.
- Provides an overview of federal and state legislation on pollution control.

BENEFITS OF A POLLUTION PREVENTION PROGRAM

In the case of pollution prevention, national environmental goals coincide with industry's economic interests. Businesses have strong incentives to reduce the toxicity and sheer volume of the waste they generate. A company with an effective, ongoing pollution prevention plan may well be the lowest-cost producer and have a significant competitive edge. The cost per unit produced will decrease as pollution prevention measures lower liability risk

A pollution prevention program addresses all types of waste.

Those companies "struggling to maintain compliance today may not be around by the end of the '90s. Those toeing the compliance line will survive. But those viewing the environment as a strategic issue will be leaders."
— Richard W. MacLean, chief of environmental programs at Arizona Public Service Co., as quoted in Environmental Business Journal, December, 1991.

and operating costs. The company's public image will also be enhanced.

Reduced Risk of Liability

You will decrease your risk of both civil and criminal liability by reducing the volume and the potential toxicity of the vapor, liquid, and solid discharges you generate. You should look at all types of waste, not just those that are currently defined as hazardous. Since toxicity definitions and regulations change, reducing the volume of wastes in all categories is a sound long-term management policy.

Environmental regulations at the federal and state levels require that facilities document the pollution prevention and recycling measures they employ for wastes defined as hazardous. Companies that produce excessive waste risk heavy fines, and their managers may be subject to fines and imprisonment if potential pollutants are mismanaged.

Civil liability is increased by generating hazardous waste and other potential pollutants. Waste handling affects public health and property values in the communities surrounding production and disposal sites. Even materials not currently covered by hazardous waste regulations may present a risk of civil litigation in the future.

Workers' compensation costs and risks are directly related to the volume of hazardous materials produced. Again, it is unwise to confine your attention to those materials specifically defined as hazardous.

Reduced Operating Costs

An effective pollution prevention program can yield cost savings that will more than offset program development and implementation costs. Cost reductions may be immediate savings that appear directly on the balance sheet or anticipated savings based on avoiding potential future costs. Cost savings are particularly noticeable when the costs resulting from the treatment, storage, or disposal of wastes are allocated to the production unit, product, or service that produces the waste. Refer to Chapter 6 for more information on allocating costs.

Materials costs can be reduced by adopting production and packaging procedures that consume fewer resources, thereby creating less waste. As wastes are reduced, the percentage of raw materials converted to finished products increases, with a proportional decrease in materials costs.

Waste management and disposal costs are an obvious and readily measured potential savings to be realized from pollution prevention. Federal and state regulations mandate special in-plant handling procedures and specific treatment and disposal methods for toxic wastes. The costs of complying with these requirements and reporting on waste disposition are direct costs to businesses. There are also indirect costs, such as higher taxes for such public

"Above all, companies want to pin down risk... Because the costs can be so enormous, risk must now be taken into account across a wide range of business decisions."
— Bill Schwalm, senior manager for environmental programs and manufacturing at Polaroid, in an interview with Environmental Business Journal, December, 1991.

Look beyond the wastes currently defined as hazardous.

A comprehensive pollution prevention program can reduce current and future operating costs.

services as landfill management. The current trend is for these costs to continue to increase at the same or higher rates. Some of these cost savings are summarized in Box 1.

Waste management costs will decrease as pollution prevention measures are implemented:

- Reduced manpower and equipment requirements for on-site pollution control and treatment
- Less waste storage space, freeing more space for production
- Less pretreatment and packaging prior to disposal
- Smaller quantities treated, with possible shift from treatment, storage, and disposal (TSD) facility to non-TSD status
- Less need to transport for disposal
- Lower waste production taxes
- Reduced paperwork and record-keeping requirements, e.g., less Toxic Release Inventory (TRI) reporting when TRI-listed chemicals are eliminated or reduced.

Box 1

Production costs can be reduced through a pollution prevention assessment. When a multi-disciplinary group examines production processes from a fresh perspective, opportunities for increasing efficiency are likely to surface that might not otherwise have been noticed. Production scheduling, material handling, inventory control, and equipment maintenance are all areas that can be optimized to reduce the production of waste of all types and also control the costs of production.

Optimizing processes and energy use reduces waste and controls production costs.

Energy costs will decrease as pollution prevention measures are implemented in various production lines. In addition, energy used to operate the overall facility can be reduced by doing a thorough assessment of how various operations interact. Chapter 8 discusses energy conservation.

Facility cleanup costs may result from a need to comply with future regulations or to prepare a production facility or off-site waste storage or disposal site for sale. These future costs can be minimized by acting now to reduce the amount of wastes of all types that you generate.

Improved Company Image

As the quality of the environment becomes an issue of greater importance to society, your company's policy and practices for controlling waste increasingly influence the attitudes of your employees and of the community at large.

Corporate image is enhanced by a demonstrated commitment to pollution prevention.

Employees are likely to feel more positive toward their company when they believe that management is committed to providing a safe work environment and is acting as a responsible member of the community. By participating in pollution prevention activities, employees can interact positively with each other and with management. Helping to implement and maintain a pollution prevention program should increase their sense of identity with company goals. This positive atmosphere helps to retain a competitive workforce and to attract high-quality new employees.

Community attitudes will be more positive toward companies that operate and publicize a thorough pollution prevention program. Most communities actively resist the siting of new waste disposal facilities in their areas. In addition, they are becoming more conscious of the monetary costs of treatment and disposal. Creating environmentally compatible products and avoiding excessive consumption and discharge of material and energy resources, rather than concentrating solely on treatment and disposal, will greatly enhance your company's image within your community and with potential customers.

Public Health and Environmental Benefits

Reducing production wastes provides upstream benefits because it reduces ecological damage due to raw material extraction and refining operations. Subsequent benefits are the reduced risk of emissions during the production process and during recycling, treatment, and disposal operations.

THE ENVIRONMENTAL MANAGEMENT HIERARCHY

The Pollution Prevention Act of 1990 reinforces the U.S. EPA's Environmental Management Options Hierarchy, which is illustrated in Figure 1. The highest priorities are assigned to preventing pollution through source reduction and reuse, or closed-loop recycling.

Preventing or recycling at the source eliminates the need for off-site recycling or treatment and disposal. Elimination of pollutants at or near the source is typically less expensive than collecting, treating, and disposing of wastes. It also presents much less risk to your workers, the community, and the environment.

WHAT IS POLLUTION PREVENTION?

Pollution prevention is the maximum feasible reduction of all wastes generated at production sites. It involves the judicious use of resources through source reduction, energy efficiency, reuse of input materials during production, and reduced water consumption. There are two general methods of source reduction that can be used in a pollution prevention program: product changes and process changes. They reduce the volume and toxicity of production wastes and of end-products during their life-cycle and at disposal. Figure 2 provides some examples.

Source reduction and reuse prevent pollution.

Change products and production processes to reduce pollution at the source.

Product changes in the composition or use of the intermediate or end products are performed by the manufacturer with the purpose of reducing waste from manufacture, use, or ultimate disposal of the products. Chapter 7 in this *Guide* provides information on designing products and packaging that have minimal environmental impact.

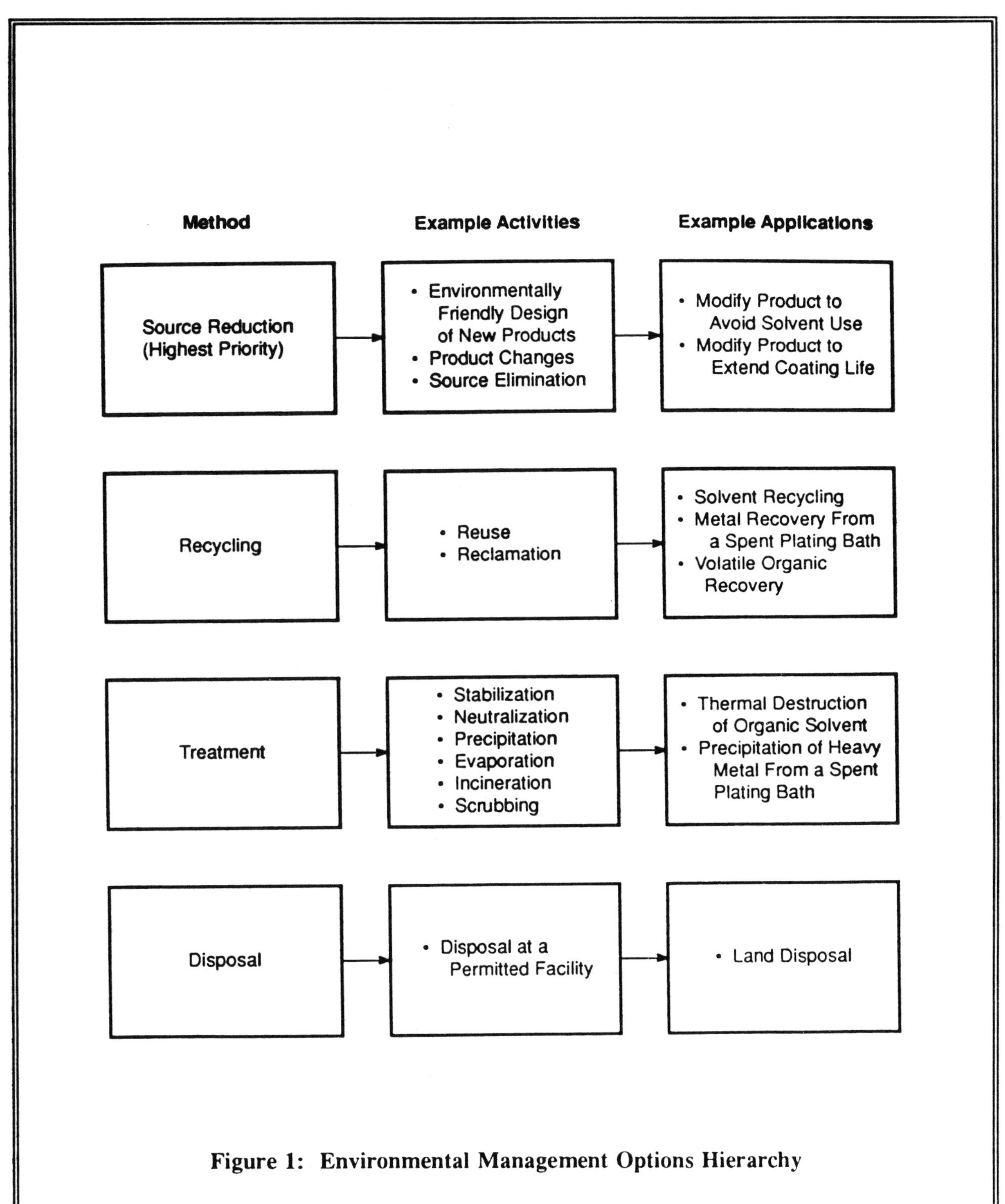

Figure 1: Environmental Management Options Hierarchy

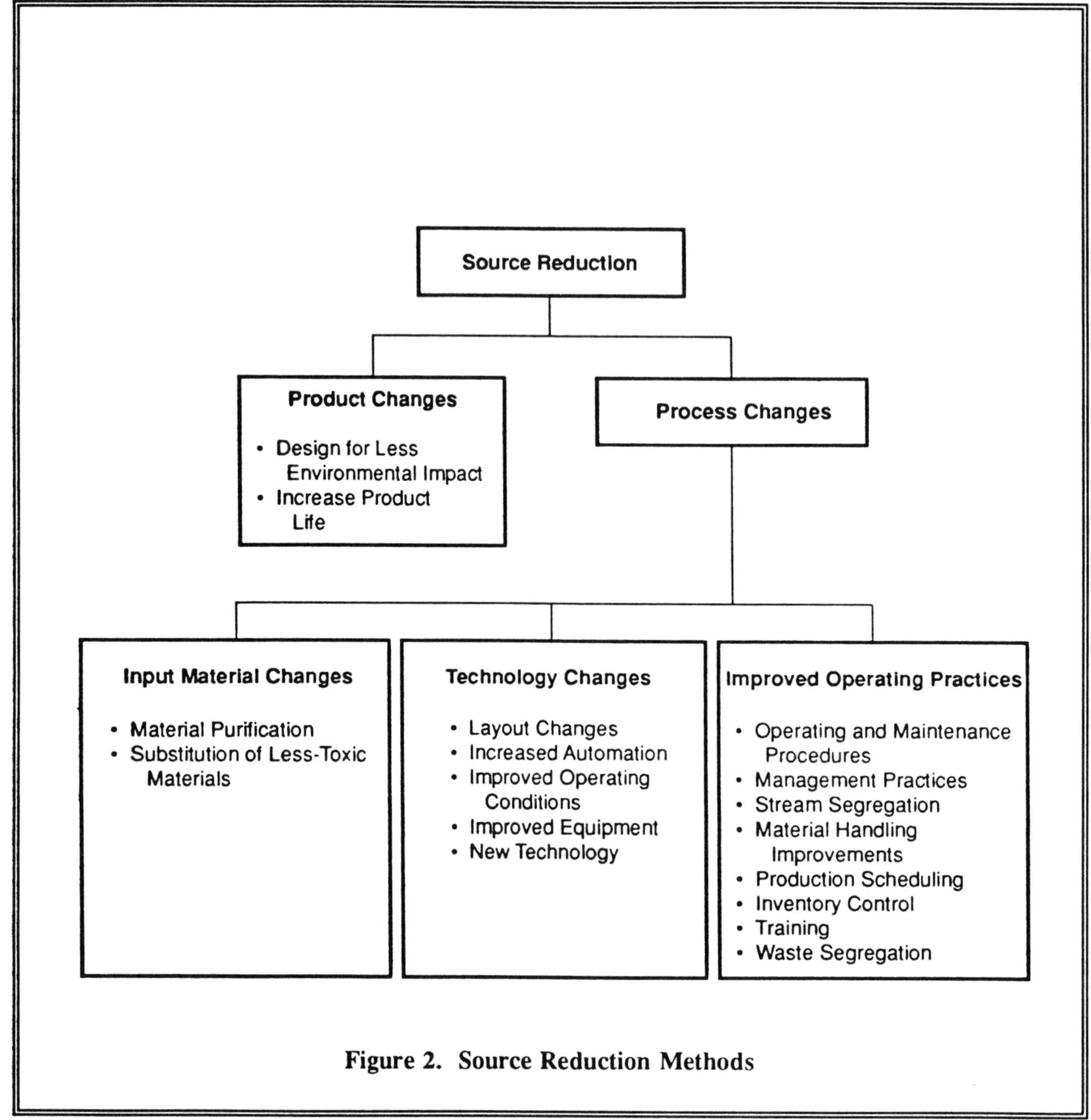

Figure 2. Source Reduction Methods

Process changes are concerned with how the product is made. They include input material changes, technology changes, and improved operating practices. All such changes reduce worker exposure to pollutants during the manufacturing process. Typically, improved operating practices can be implemented more quickly and at less expense than input material and technology changes. Box 2 provides examples of process changes.

Process changes may be implemented more quickly than product changes.

The following process changes are pollution prevention measures because they reduce the amount of waste created during production.

Examples of input material changes:
- Stop using heavy metal pigment.
- Use a less hazardous or toxic solvent for cleaning or as coating.
- Purchase raw materials that are free of trace quantities of hazardous or toxic impurities.

Examples of technology changes:
- Redesign equipment and piping to reduce the volume of material contained, cutting losses during batch or color changes or when equipment is drained for maintenance or cleaning.
- Change to mechanical stripping/cleaning devices to avoid solvent use.
- Change to a powder-coating system.
- Install a hard-piped vapor recovery system to capture and return vaporous emissions.
- Use more efficient motors.
- Install speed control on pump motors to reduce energy consumption.

Examples of improved operating practices:
- Train operators.
- Cover solvent tanks when not in use.
- Segregate waste streams to avoid cross-contaminating hazardous and nonhazardous materials.
- Improve control of operating conditions (e.g., flow rate, temperature, pressure, residence time, stoichiometry).
- Improve maintenance scheduling, record keeping, or procedures to increase efficiency.
- Optimize purchasing and inventory maintenance methods for input materials. Purchasing in quantity can reduce costs and packaging material if care is taken to ensure that materials do not exceed their shelf life. Reevaluate shelf life characteristics to avoid unnecessary disposal of stable items.
- Stop leaks, drips, and spills.
- Turn off electrical equipment such as lights and copiers when not in use.
- Place equipment so as to minimize spills and losses during transport of parts or materials.
- Use drip pans and splash guards.

Box 2

WHAT IS NOT POLLUTION PREVENTION?

There are a number of pollution control measures that are applied only after wastes are generated. They are, therefore, not correctly categorized as pollution prevention. Box 3 provides some examples of procedures that are waste handling, not pollution prevention, measures.

Waste treatment is not pollution prevention.

The following are not pollution prevention measures because they are taken after the waste is created:

- **Off-site recycling:**
  Off-site recycling (e.g., solvent recovery at a central distillation facility) is an excellent waste management option. However, it does create pollution during transport and during the recycling procedure.

- **Waste treatment:**
  Waste treatment involves changing the form or composition of a waste stream through controlled reactions to reduce or eliminate the amount of pollutant. Examples include detoxification, incineration, decomposition, stabilization, and solidification or encapsulation.

- **Concentrating hazardous or toxic constituents to reduce volume:**
  Volume reduction operations, such as dewatering, are useful treatment approaches, but they do not prevent the creation of pollutants. For example, pressure filtration and drying of a heavy metal waste sludge prior to disposal decreases the sludge water content and waste volume, but it does not decrease the number of heavy metal molecules in the sludge.

- **Diluting constituents to reduce hazard or toxicity:**
  Dilution is applied to a waste stream after generation and does not reduce the absolute amount of hazardous constituents entering the environment.

- **Transferring hazardous or toxic constituents from one environmental medium to another:**
  Many waste management, treatment, and control practices used to date have simply collected pollutants and moved them from one environmental medium (air, water, or land) to another. An example is scrubbing to remove sulfur compounds from combustion process off-gas.

Box 3

Off-site recycling is vastly preferable to other forms of waste handling because it helps to preserve raw materials and reduces the amount of material that will require disposal. However, compared with closed-loop recycling (or reuse), performed at the production site, there is likely to be more residual waste that will require disposal. Further, waste transportation and the recycling process itself carry the risks of worker exposure and of release into the environment.

Off-site recycling carries some risk.

Transferring hazardous wastes to another environmental medium is not pollution prevention. Many waste management practices to date have simply collected pollutants and moved them from one environmental medium to another. For example, solvents can be removed from wastewater by means of an activated carbon

Transfer to another environmental medium should be avoided in most cases.

adsorbers. However, regenerating the carbon requires the use of another solvent or heating, which transfer the waste to the atmosphere. In some cases, transfer is a valid treatment option. However, too often the purpose has been to shift a pollutant to a less-tightly regulated medium. In either case, media transfers are not pollution prevention.

Waste treatment prior to disposal reduces the toxicity and/or disposal-site space requirements but does not eliminate all pollutant materials. This includes such processes as volume reduction, dilution, detoxification, incineration, decomposition, stabilization, and isolation measures such as encapsulation or embedding.

POLLUTION PREVENTION REGULATORY FRAMEWORK

Companies are required to have pollution prevention programs for waste classified as hazardous. See Appendix D for points of contact at U.S. and state agencies levels who can provide you with information about regulations and with technical assistance for pollution prevention.

Hazardous waste reduction programs are required under RCRA, PPA, and CERCLA.

Federal

Under the terms of the 1988 **Resource Conservation and Recovery Act (RCRA)**, "it shall be a condition of any permit issued under this section for the treatment, storage, or disposal of hazardous waste on the premises where such waste was generated that the permittee certify, no less often than annually, that the generator of the hazardous waste has a program in place to reduce the volume or quantity and toxicity of such waste to the degree determined by the generator to be economically practicable."

The 1990 **Pollution Prevention Act (PPA)** specifies that facilities required to report releases to the U.S. EPA for the **Toxic Release Inventory (TRI)** provide documentation of their procedures for preventing the release of or for reusing these materials (Box 4).

These acts, plus the **Comprehensive Environmental Response, Compensation, and Liability Act (CERCLA)**, require generators of hazardous wastes to evaluate and document their procedures for controlling the environmental impact of their operations.

However, the **PPA** goes beyond wastes designated as hazardous. It encourages the maximum possible elimination of wastes of all types. It emphasizes that the preferred method of preventing pollution is to reduce at the source the volume of waste generated and that reuse (closed-loop recycling) should be performed whenever possible. In this way, it is fundamentally different from off-site recycling, treatment, and disposal and is meant to reduce the need for these measures. Treatment and disposal are to be viewed as last-resort measures.

The Pollution Prevention Act encourages source reduction of all waste types.

Pollution Prevention Act of 1990 data reporting requirements for TRI chemicals:

- Amount entering any waste stream (or otherwise released into the environment) before recycling, treatment, or disposal, and the percent change from the previous year.
- Amount recycled on site or off site during each calendar year, the percent change for the previous year, and the recycling process used.
- Source reduction practices used during each year.
- Amount expected to be reported under the first two data items above for the two calendar years right after the reporting year (reported as percent change).
- Ratio of reporting year's production to previous year's production.
- Techniques used to identify source reduction opportunities.
- Amount released into the environment from a catastrophic event, remedial action, or other one-time event and not associated with the production process.

Box 4

State

A number of states have enacted legislation that requires pollution prevention or waste minimization. As of March, 1992, a total of 26 states had passed such legislation (WRITAR *Survey of State Legislation*, March 1992). (See Box 5.)

State legislation, if enacted, must address at a minimum those substances defined as hazardous by RCRA, CERCLA, and the Superfund Amendments and Reauthorization Act of 1986 (SARA). Additional substances may be classified as hazardous by the individual state. Most programs are aimed at large-quantity generators since they are the high-volume producers of pollution. Some also apply to small-quantity generators or have special provisions for these. Fifteen states require waste generators to submit plans and/or progress reports on waste minimization or pollution prevention efforts, while others make such reporting optional.

In many states, the legislation establishes pollution prevention program offices, advisory boards, or commissions to provide technical assistance and to promote education, training, and research.

Some states require pollution prevention programs.

State legislation promoting pollution prevention as of March, 1992:

Alaska	Solid and Hazardous Waste Management Act
Arizona	Amendments to Arizona Hazardous Waste Management Statutes
California	Hazardous Waste Reduction and Management Review Act
Connecticut	Environmental Assistance to Business Act
Delaware	Waste Minimization/Pollution Prevention Act
Florida	Pollution Prevention Act
Georgia	Amendment to Hazardous Waste Management Act
Illinois	Toxic Pollution Prevention Act
Indiana	Amendment to Environmental Code
Iowa	Toxics Pollution Prevention Act
Kentucky	(no title)
Louisiana	Waste Reduction Law
Maine	Reduction of Toxics Use, Waste and Release Act
Massachusetts	Toxic Use Reduction Act
Minnesota	Toxic Pollution Prevention Act
Mississippi	Comprehensive Multimedia Waste Minimization Act
New Jersey	Pollution Prevention Act
New York	Hazardous Waste Management Act
North Carolina	Hazardous Waste Management Act
Oregon	Toxic Use Reduction and Hazardous Waste Reduction Act
Rhode Island	Hazardous Waste Facility Planning Act
Tennessee	Hazardous Waste Reduction Act
Texas	Waste Reduction Policy Act
Vermont	Hazardous Waste Management Act
Washington	Hazardous Waste and Substance Reduction Act
Wisconsin	Hazardous Substances, Toxic Pollutants, Hazardous Waste Use and Release Reduction

Colorado, Michigan, Missouri, Ohio, and South Carolina are expected to enact pollution prevention regulations in 1992.

Box 5

Pollution prevention planning is a comprehensive and continual evaluation of how you do business, and the resulting program will affect many functional areas within your company. Therefore, it has much in common with the planning you already do for other aspects of your business operations.

Figure 3 illustrates the major steps in the pollution prevention program. These steps are described in this chapter and in Chapters 3 through 5.

This chapter considers the elements of pollution prevention program design as they might be addressed by a small- or medium-sized company. These elements include building support for pollution prevention throughout the company, organizing the program, setting goals and objectives, performing a preliminary assessment of pollution prevention opportunities, and identifying potential problems and their solutions.

Pollution prevention should be integrated into your overall business plan.

ESTABLISH THE POLLUTION PREVENTION PROGRAM

Executive Level Decision

In some companies, the initiative to investigate setting up a pollution prevention program will be taken at the executive level. In others, lower-level managers or employees will be the catalysts. In either case, it may be necessary to gather information to demonstrate that pollution prevention opportunities exist and should be explored. This information will be used by company executives as they weigh the potential value of pollution prevention and decide whether to commit the resources necessary to develop and implement the program.

One way to gather this information is to perform a preliminary assessment. A pre-assessment is part of the formal program design effort and is, therefore, described later in this chapter. However, a high-level pre-assessment of only one or two areas of the facility can be done to gather information and, perhaps, even identify several low-cost, quick-payoff pollution prevention techniques that can be implemented readily.

Once senior managers have decided to establish a pollution prevention program, they should convey this commitment to all employees through a formal policy statement. This will establish a framework for communicating the formal commitment throughout the organization.

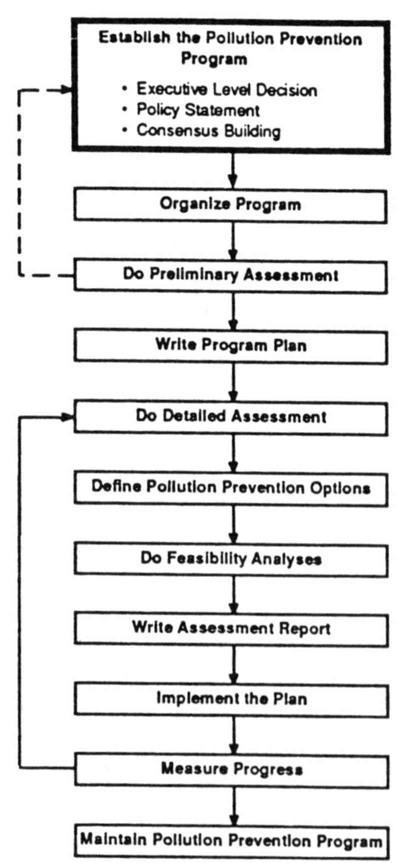

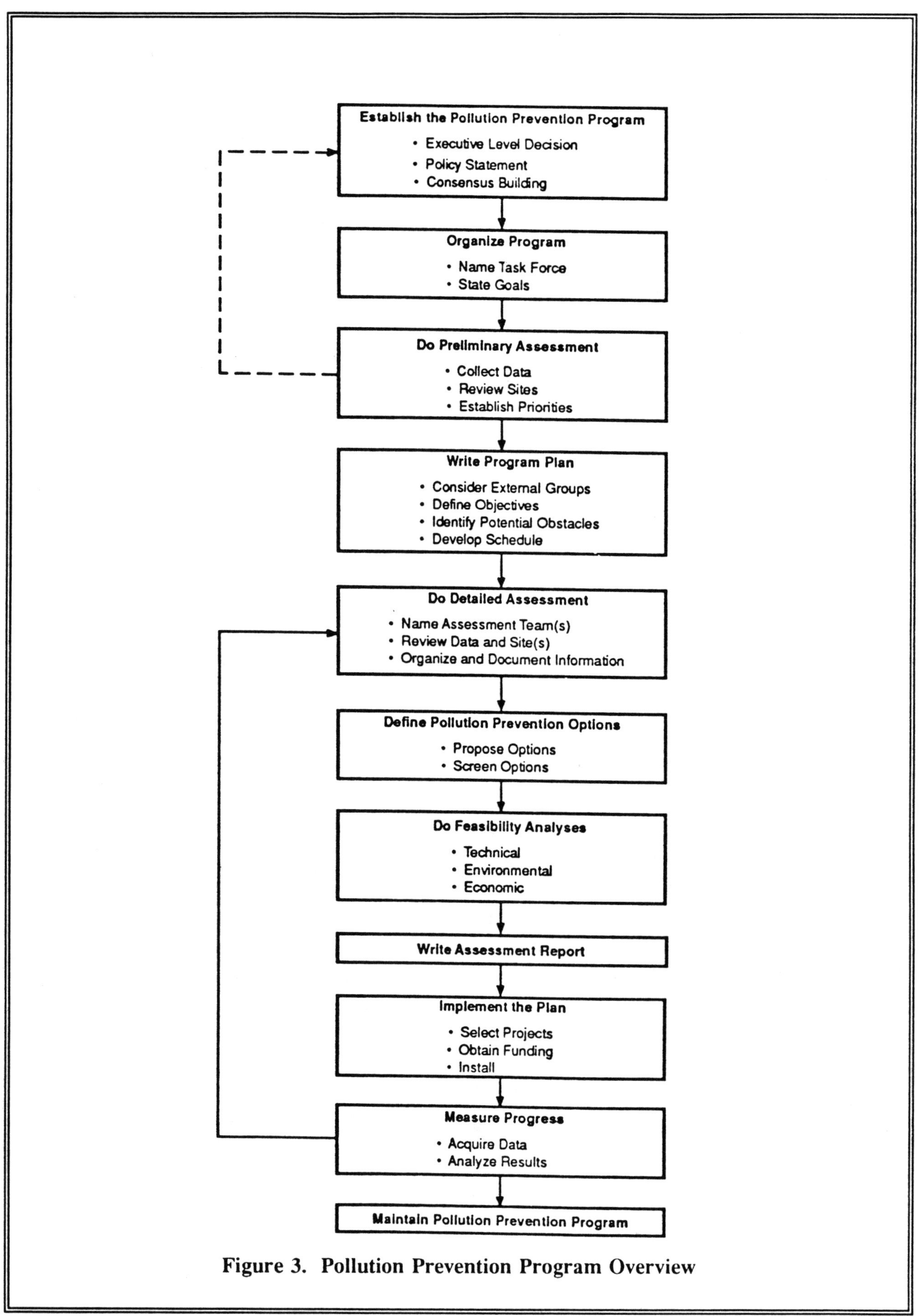

Figure 3. Pollution Prevention Program Overview

Policy Statement

As with other policy statements your company develops, your pollution prevention policy statement should state why a program is being established, what is to be accomplished in qualitative terms, and who will do it. Two example policy statements are given in Box 6. They differ in level of detail, but both answer these key questions:

The policy statement is the foundation of the pollution prevention program.

Why are we implementing pollution prevention?
>   *We want to protect the environment.*

What will be done to implement pollution prevention?
>   *We will reduce or eliminate the amounts of all types of waste, and we will improve energy efficiency.*

Who will implement pollution prevention?
>   *Everyone will be involved.*

Consensus Building

After you have developed your pollution prevention policy statement, consider how it should be presented to your employees so that they will see it as an ongoing, company-wide commitment.

It is essential that employees understand and support the pollution prevention program.

Everyone in your facility will be involved in some way.

- **POLICY STATEMENT EXAMPLE 1** — "(Your Company Name) is committed to excellence and leadership in protecting the environment. In keeping with this policy, our objective is to reduce waste and emissions. We strive to minimize adverse impact on the air, water, and land through pollution prevention and energy conservation. By successfully preventing pollution at its source, we can achieve cost savings, increase operational efficiencies, improve the quality of our products and services, maintain a safe and healthy workplace for our employees, and improve the environment. (Your Company Name)'s environmental guidelines include the following:

  - Environmental protection is everyone's responsibility. It is valued and displays commitment to (Your Company Name).

  - We will commit to including pollution prevention and energy conservation in the design of all new products and services.

  - Preventing pollution by reducing and eliminating the generation of waste and emissions at the source is a prime consideration in research, process design, and plant operations. (Your Company Name) is committed to identifying and implementing pollution prevention opportunities through encouraging and involving all employees.

  - Technologies and methods which substitute nonhazardous materials and utilize other source reduction approaches will be given top priority in addressing all environmental issues.

  - (Your Company Name) seeks to demonstrate its responsible corporate citizenship by adhering to all environmental regulations. We promote cooperation and coordination between industry, government, and the public toward the shared goal of preventing pollution at its source."

- **POLICY STATEMENT EXAMPLE 2** — "At (Your Company Name), protecting the environment is a high priority. We are pledged to eliminate or reduce our use of toxic substances and to minimize our use of energy and generation of all wastes, whenever possible. Prevention of pollution at the source is the preferred alternative. When waste cannot be avoided, we are committed to recycling, treatment, and disposal in ways that minimize undesirable effects on air, water, and land."

(Adapted from: Waste Reduction Institute for Training and Applications Research, Inc. [WRITAR], *Survey and Summaries*, 1991, and Minnesota Office of Waste Management, Feb. 1991, *Minnesota Guide to Pollution Prevention Planning*)

Box 6

While executives and managers will assign priorities and set the tone for the pollution prevention program, the attitude of production-level employees will have a significant effect on its success. Since it is their daily activities that generate waste, their support of the program is essential.

How you publicize the policy depends on the size and the culture of your company. You may decide to call a general meeting or to hold several meetings with smaller groups. There may be other types of publicity that you have found effective.

Encourage employee participation.

You might offer bonuses or other awards to employees who suggest ways to prevent pollution. Announcing awards in newsletters or on bulletin boards provides additional incentive to employees and further publicizes the program. Pollution prevention might be included in job objectives and performance evaluations for managers and other appropriate employees.

In any case, it is important to emphasize your company's commitment to pollution prevention and encourage employee participation. This will help to establish a positive atmosphere and reassure employees who might be concerned about the changes that will result. This approach will also elicit worthwhile pollution prevention suggestions.

A positive atmosphere produces best results.

Employees feel committed to pollution prevention when they are encouraged to:

- Help define company goals and objectives.
- Review processes and operations to determine where and how toxic substances are used and hazardous wastes are generated.
- Recommend ways to eliminate or reduce waste production at the source.
- Design or modify forms and records to monitor materials used and waste.
- Find ways to involve suppliers and customers.
- Think of ways to acknowledge and reward employee contributions to the pollution prevention effort.

Box 7

ORGANIZE THE POLLUTION PREVENTION PROGRAM

The program will be directed by the Pollution Prevention Task Force. Their first task will be to delineate program goals.

Name the Pollution Prevention Task Force

The people who will direct the pollution prevention program should be selected carefully. They will have overall responsibility for developing the plan and directing its implementation. Their capabilities and their attitudes toward the effort will be major determinants of how successful it is. As with other areas of your operation, successful program execution will require integration and continuity of the planning, implementation, modification, and maintenance stages. Therefore, all individuals named to this task force should have substantial technical, business, and communication skills as well as thorough knowledge of the company. The responsibility and authority of each individual should be established during this organizational stage.

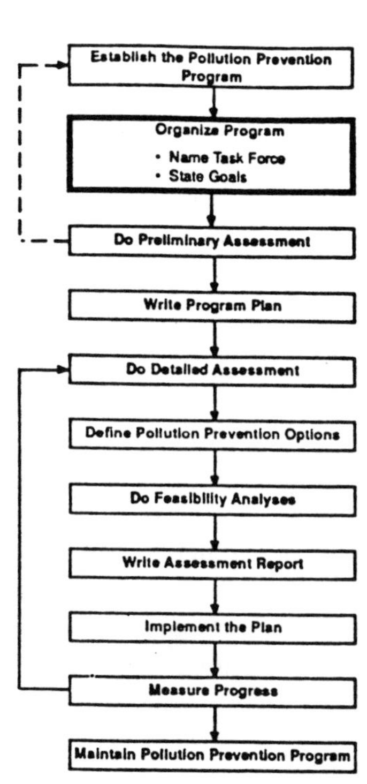

The **program leader** should be named from the highest level practical. The leader must have the authority and the influence necessary to keep the program on track and to ensure that pollution prevention becomes an integral part of the overall corporate plan. The role of the leader is to facilitate the flow of information among all levels in the company. Therefore, the leader should possess the personal qualities necessary to elicit broad-based support from the company's employees.

The task force works together during planning and preassessment.

One or more **pollution prevention champions** should be designated. The task of a "champion" is to overcome possible resistance to proposed changes in operations. In a medium-sized company, several champions may be assigned, perhaps according to production area. In a very small company, the champion may also be the program leader. Champions will be the team members who are the most visible within the production areas and should be respected and trusted at all levels in order to perform this liaison role well.

Other team members might be selected for their specific technical or business expertise. Environmental and plant process engineers, production supervisors, and experienced line-workers are good candidates. Other potential sources include purchasing and quality-assurance staff. In some cases, outside consultants may be retained to work with the in-house team.

Once the task force has been established, they will be a valuable resource within the company. When plans are being made to

The task force will direct the development and implementation of the pollution prevention program and help integrate its principles into all phases of corporate planning.

expand the facility or to design or redesign products, they can review the plans to determine whether waste generation has been evaluated thoroughly.

State Goals

The program leaders will need to establish goals that state the long-term direction for the pollution prevention program. Well-defined goals will help to focus effort and build consensus. Goals should be consistent with your company's pollution prevention policy and, in fact, may have been stated in general terms in the policy statement. Now, they need to be stated more specifically.

The goal-setting process will involve the program team and company management. The size of the group needed to develop the goals depends on the size and complexity of your facility. For a small company, the group might be only two or three people.

Since success in pollution prevention may require basic changes in the corporate culture, goals should be useful and meaningful for every employee. Goals need to be challenging enough to motivate but not unreasonable or impractical.

When beginning the goal-setting process, consider starting from the zero-discharge perspective. This ideal situation would involve 100% utilization of resources, eliminating disposal costs and regulatory compliance needs. This is probably not a completely achievable goal in any industry, given current technology. However, like zero-defect production goals, zero-discharge goals encourage an attitude of continually striving for improvement.

Pollution prevention goals can be qualitative, such as, "achieve a significant reduction of toxic substance emissions to the environment." Quantitative goals are more difficult to develop but are worth the extra effort. They spell out your pollution prevention commitment and give all participants and observers a yardstick for measuring progress.

Finally, goals should be flexible and adaptable. Conditions change in actual practice. As your pollution prevention program becomes more focused and the pollution-specific aspects of the operation become better known, the goals can be refined. They can be adjusted up or down as the program matures and lessons are learned. Periodic goal-achievement review and adjustment will keep your program active and visible within the company.

Your corporate pollution prevention policy and goals should be integrated in a formal planning document.

DO THE PRELIMINARY ASSESSMENT

Even though you may have completed some aspects of the preliminary assessment as input to the executive decision to develop a pollution prevention program, a deeper examination will be needed at this point. The data collection that is a part of this pre-assessment will help the team review the data that are already available and begin defining ways to process that data. These data and the site visits will enable the Task Force to establish priorities and procedures for detailed assessments. Chapter 3 describes the detailed assessment phase and the more in-depth data collection and analyses that will be done at that stage.

Collect Data

The extent and complexity of the system for collecting pollution prevention data should be consistent with the needs of your company. Keep in mind that the goal of the program is to prevent pollution, not to collect data — the simplest system that fits your needs is the best. Depending on the nature and size of your firm, much of the data needed for a pollution prevention program may be collected as a normal part of plant operations or in response to existing regulatory requirements. (See Box 8.) The worksheets in Appendix A can be used for the pre-assessment; you may decide to modify them to fit your particular industry.

An all-media approach, which deals with all air, water, and solid waste emissions and releases, will be the most effective. This involves considering all waste streams, identifying their sources and quantifying the true costs of pollution control, treatment, and waste disposal. There are a number of information sources to consider.

Regulatory reports — National Pollutant Discharge Elimination System (NPDES) and SARA Title III reports document the volume, composition, and degree of toxicity of wastewater discharged. The toxic substance release inventories required by SARA Title III, Section 313 may provide information on emissions into all environmental media.

Engineering and operating data — Shipping manifests will provide quantities of hazardous waste shipped during a given period, but may lack chemical analysis, specific source, and the time period during which the waste was generated. The plant design documents and equipment operating manuals and procedures may yield specific data for streams inside of the plant.

Plant business records — Records available from inventory control, purchasing, records management, accounting, marketing, and training can provide data needed for the pre-assessment and may themselves present opportunities for pollution prevention. For example, improved inventory control and judicious purchasing can significantly reduce the volume of raw materials that must be disposed of because they become outdated. In reviewing existing data, you may find that current accounting practices are not appropriate for placing the burden of pollution and pollution control at the point of generation. These findings should be taken into account when costs of pollution control measures are analyzed. (See Chapter 6.)

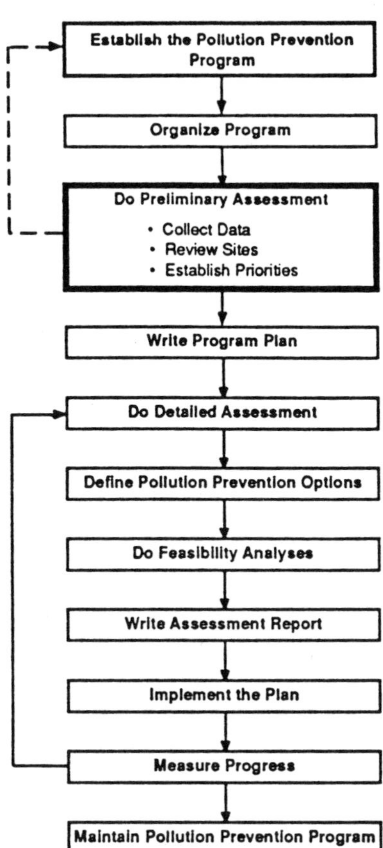

Review existing information resources.

**Data sources for facility
information include:**

Regulatory Information:
- Waste shipment manifests
- Emission inventories
- Biennial hazardous waste reports
- Waste, wastewater, and air emissions analyses, including intermediate streams
- Environmental audit reports
- Permits and/or permit applications
- Form R for SARA Title III Section 313

Process Information:
- Process flow diagrams
- Design and actual material and heat balances for:
  - production processes
  - pollution control processes
- Operating manuals and process descriptions
- Equipment lists
- Equipment specifications and data sheets
- Piping and instrument diagrams
- Plot and elevation plans
- Equipment layouts and logistics

Raw Material/Production Information:
- Product composition and batch sheets
- Material application diagrams
- Material safety data sheets
- Product and raw material inventory records
- Operator data logs
- Operating procedures
- Production schedules

Accounting Information:
- Waste handling, treatment, and disposal costs
- Water and sewer costs, including surcharges
- Costs for nonhazardous waste disposal, such as trash and scrap metal
- Product, energy, and raw material costs
- Operating and maintenance costs
- Department cost accounting reports

Other Information:
- Environmental policy statements
- Standard procedures
- Organization charts

Box 8

Visit Sites

In order to utilize resources of time, staff, and money wisely, the task force will need to prioritize the processes, operations, and wastes that will be addressed during the subsequent detailed assessment phase. During that phase, they will target the most important waste problems, moving on to lower-priority problems as resources permit. The pre-assessment site visits will provide the information needed to accomplish this prioritization and to designate the detailed assessment teams, who will be selected for their expertise in particular areas.

Site visits make it possible to:
- *prioritize areas*
- *select detailed assessment teams*

Establish Priorities

Assigning priorities (Box 9) to processes, operations, and materials will focus the remainder of the pollution prevention plan development effort. The priorities set in this stage will guide the selection of areas for the detailed assessments. Areas may also be targeted based on the volume of waste produced or the cost of waste disposal. Regulatory concerns such as the RCRA land disposal restrictions or SARA Title 313 chemicals may also guide prioritization. The Option Rating Weighted Sum Method, which is illustrated in Appendix E, can be used during the pre-assessment phase as well as during detailed assessment.

The priorities established at this point will guide subsequent effort.

PREPARE THE PROGRAM PLAN

With the information collected during the pre-assessment, the Task Force can develop a detailed program plan. This plan will address the extent to which external organizations will be involved, define pollution prevention program objectives, identify potential obstacles and solutions, and define the data collection and analysis procedures that will be used. A summary of the points that should be addressed in a program plan appears in Box 10.

Contacting External Groups

At this point, the Task Force should consider soliciting input from outside the company. Including the surrounding community in the pollution prevention planning process can create a new

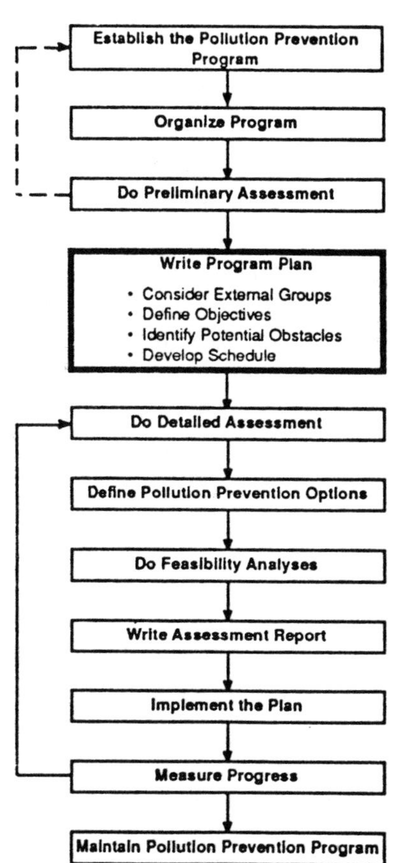

forum for communication. Valuable technical information can also be exchanged with some organizations.

The formal written pollution prevention plan will include the following elements:

- Corporate policy statement of support for pollution prevention
- Description of your pollution prevention planning team(s) makeup, authority, and responsibility
- Description of how all of the groups (production, laboratory, maintenance, shipping, marketing, engineering, and others) will work together to reduce waste production and energy consumption
- Plan for publicizing and gaining company-wide support for the pollution prevention program
- Plan for communicating the successes and failures of pollution prevention programs within your company
- Description of the processes that produce, use, or release hazardous or toxic materials, including clear definition of the amounts and types of substances, materials, and products under consideration
- List of treatment, disposal, and recycling facilities and transporters currently used
- Preliminary review of the cost of pollution control and waste disposal
- Description of current and past pollution prevention activities at your facility
- Evaluation of the effectiveness of past and ongoing pollution prevention activities
- Criteria for prioritizing candidate facilities, processes, and streams for pollution prevention projects.

Box 10

Legislative and executive officials can provide their perspectives on environmental protection issues and information on their planning processes. In return, they can gain information that will help them make decisions on future public issues related to the environment.

Communication with government and community leaders yields mutual benefits.

Community involvement is a good way to build credibility and focus pollution prevention efforts on the discharge paths that most concern your neighbors. However, it may be wise to wait until the program is established before seeking to involve the community. Having a few pollution prevention projects underway will demonstrate your good faith. Positive community involvement can be encouraged through holding open meetings, granting interviews to the media, advertising, direct-mail surveys and opinion polls.

Other businesses can be a source of information on technical issues and suppliers, either because they are in the same geographical area or because they have similar technical areas of interest. Local business groups are a good way of locating resources in the immediate area, while trade and professional associations can provide contacts in other parts of the country or the world. Of course, the companies with the most similar interests may be

Other businesses will have useful information.

competitors, but it should be possible to interact without risking disclosure of business-sensitive information.

Define Objectives

During the preliminary assessment phase, the program team will have identified opportunities for pollution prevention and will have worked with the executive group to establish priorities. These will be the starting point for defining short- and long-range objectives.

Objectives are the specific tasks that will be necessary to achieve goals. For example, in order to reach a goal of reducing waste, the objectives might be defined as reducing solvent, paper, and packaging wastes by specific amounts over a stated period of time.

Objectives can be defined at the facility- or the department-level, depending on the size and diversity of your company. A small company could decide to develop a single set of objectives to cover all of its operations. A larger company with many facilities or products might develop an overall corporate plan describing goals and objectives, supplemented by facility- or product-specific goals. In any case, the management at each location must understand and support its objectives if the pollution prevention program is to be successful.

Objectives should be stated in quantitative terms and should have target dates. These two attributes make objectives effective tools for directing effort and measuring progress.

Identify Potential Obstacles

As the pollution prevention program team begins to develop and implement a pollution prevention program, they are likely to encounter a number of factors that will complicate the process. These need to be recognized, and the means for overcoming them need to be defined. Apparent obstacles will be less likely to impede the process if everyone understands that there is a mechanism for addressing them in a later stage.

The mix of factors and the relative degree of difficulty each presents will vary from company to company. Those that are likely to be encountered by most businesses are discussed below. They fall into four broad categories: **economic, technical, regulatory,** and **institutional**.

Economic Obstacles. The task force should recognize that some complex economic factors may need to be addressed later. Broadly defining procedures now for dealing with them will help prevent economic concerns from stifling the creative process of defining options.

Cost-benefit analysis procedures should be defined. Many proposed pollution prevention options will have start-up costs. For example, additional or replacement equipment may need to be purchased, staff training may be required, or alternative raw materials may cost more. Some of these additional costs can be justified readily because they clearly will be cost-effective and will have short pay-back times. However, many will not be so clearcut and will need more sophisticated analysis. Chapter 6 describes the "Total Cost Assessment" (TCA) approach as it applies to pollution prevention projects and discusses why it may be necessary to look at longer payback times for pollution prevention projects.

Limited financial resources for capital improvements may also be a problem, even for options that will ultimately be profitable. The team should investigate the availability of and conditions for funding assistance or low-interest loans from state or local agencies. Appendix D provides information on whom to contact.

Technical Obstacles. Information will be needed on alternative procedures that should be considered, how to integrate them in the production process, and what side effects are possible.

Information resources could be a problem. As a small or medium-sized business, you may not have ready access to a central source of information on pollution prevention techniques. There are several ways to deal with this problem. Contact appropriate agencies listed in Appendix D for assistance. Encourage employees to watch for information in the technical journals and newsletters they read and to pass it on to the task force. Those who belong to professional societies may get ideas from other members. Metropolitan or university library reference departments can provide assistance in locating sources of published information as well as names of people who might be able to provide information in specific areas. If the scope of the technical problem and resources permits, it may be appropriate to retain a consultant.

Limited flexibility in the manufacturing process may pose another technical barrier. A proposed pollution prevention option may involve modifying the work flow or the product or installing new equipment; implementation could require a production shutdown, with loss of production time. You might be concerned that the new operation will not work as expected or might create a bottleneck that slows production. In addition, the production facility might not have space for pollution prevention equipment. These technical barriers can be overcome by having design and production personnel take part in the planning process and by using tested technology or setting up pilot operations.

Product quality or customer acceptance concerns might cause resistance to change. For example, in some printing and publishing operations it is possible to minimize waste by substituting a water-based ink for a solvent-based ink. But for some products, quality suffers when water-based ink is used. You should plan to avoid potential product quality degradation by verifying customer needs, testing the new process or product, and increasing quality control during manufacture.

There are a number of sources of technical assistance:

- **Trade associations** generally provide assistance and information about environmental regulations and various available techniques for complying with these regulations. Their information is especially valuable because it is tailored to the specific industry.
- **Published literature** can be a valuable resource. Articles in technical magazines, trade journals, government reports, and research briefs describe pollution prevention technologies and applications.
- **Federal, state, and local environmental agencies** are expanding their pollution prevention technical assistance programs. These programs make available information on industry-specific pollution prevention techniques. (See Appendix D for addresses and phone numbers of such resources.)
- **Equipment vendors** and sales literature are helpful in identifying and analyzing potential equipment-oriented options.
- **Consultants** — Consultants with experience in pollution prevention in the specific industry can usually be located.
- **Other Companies**.

Box 11

Regulatory Obstacles. Regulations may be a barrier to some pollution prevention options. For example, changing to another feed material may require changing the existing permits. In addition, it may be necessary to learn what regulations might apply to proposed alternative input materials.

Working with regulatory bodies will help resolve questions as to requirements that pertain to proposed changes.

Working with the appropriate regulatory bodies early in the planning process will help overcome this barrier. The U.S. EPA and the state environmental agencies have developed a number of documents to facilitate pollution prevention efforts by industry; some are listed in Appendix G. Points of contact at the appropriate agencies will be helpful; many are listed in Appendix D.

Your local health department and city and county waste disposal and treatment offices can also provide assistance. Industry task forces and consultants might also be contacted.

Institutional Obstacles. As with any other new program, general resistance to change and friction among elements within the organization may arise. These can result from many factors, such as lack of awareness of corporate goals and objectives, individual or organizational resistance to change, lack of commitment, poor internal communication, requirements of existing labor contracts, or an inflexible organizational structure.

Resistance to change and friction among organizational elements can be reduced by effective communication.

Analyze these barriers from different perspectives in order to understand the concerns. Management is concerned with production costs, efficiency, productivity, return on investment, and present and future liability. Workers are concerned about job security, pay, and workplace health and safety. The extent to which these issues are addressed in the pollution prevention program will affect the success of the program.

Institutional barriers can be overcome with education and outreach programs. As was pointed out earlier, it is vital to gain the support of staff at all levels very early in the pollution prevention effort.

Develop Schedule

The final aspect of planning your pollution prevention program is to list the milestones within each of the stages from detailed assessment through implementation and assign realistic target dates. The execution of these stages (described in Chapter 3) should follow this schedule closely. Significant deviations may cause the program to falter because certain steps are not completed. Adherence to the schedule will also help control the startup or implementation costs of the program.

DEVELOPING AND IMPLEMENTING POLLUTION PREVENTION PROJECTS

This chapter outlines how to execute the pollution prevention program plan that resulted from the activities outlined in Chapter 2. The figure to the right illustrates the steps that will be discussed in this chapter and places them in the context of the overall effort.

As with the other stages, the degree of formality should be tailored to the size of the company and the diversity of its product lines. Thus, a small company may need to do only one detailed assessment and prepare one implementation plan, while a larger, more diverse company might require several in order to address all production processes. If multiple plans are developed, it will be necessary to examine how they fit together, resolving any conflicts and prioritizing them to fit available resources.

DETAILED ASSESSMENT PHASE

As part of your program design, you probably did a preliminary assessment of your facility to identify areas of opportunity for pollution prevention. Now, detailed assessments will focus on specific areas targeted by the preliminary assessment.

Assessment teams will be assigned to each operational area of the facility to gather data for later analysis. As was the case during the preliminary assessment, they will use existing written materials and site evaluations. However, they will delve much more deeply into each production process, interviewing workers and compiling necessary data that may not have been collected before.

During this process, the team may identify some options that can be implemented quickly and with little cost or risk. It is likely, however, that many options will be more complex and will require in-depth analysis later.

Designate the Detailed Assessment Team(s)

The detailed assessment phase should be started by a member of the pollution prevention task force, which was identified during program design. Unless your company is small enough that the task force and the detailed assessment team are the same, you will need to name additional staff to comprise one or more detailed assessment teams. The focus of each assessment team will be relatively specific. It is likely that three to six people will prove to

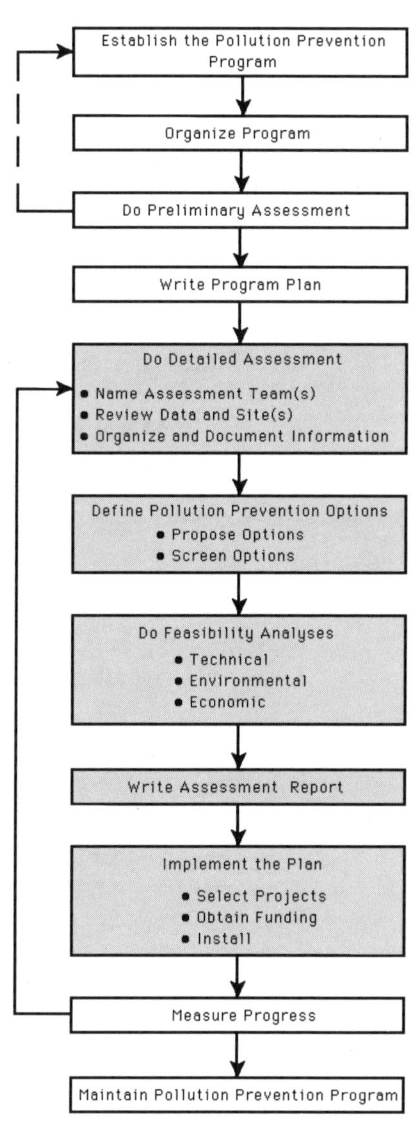

be a workable number for an assessment team. Specialists can be consulted as needed. Ideally, one member of the task force will be included on each team; this will facilitate communication. The additional team members should be people with direct responsibility for and knowledge of the waste streams and/or areas of the facility under consideration. A multidisciplinary team is likely to be more successful in achieving a comprehensive assessment and providing the best input possible to the data analysis and option definition stages. To the extent practical, you should consider engineers, supervisors, and production workers as well as finance and accounting, purchasing, and administrative staff when selecting the team members.

Aside from field of expertise, consider a candidate's ability to work on a team, apparent interest in and commitment to the program, and capacity for looking at situations from new perspectives and for thinking creatively.

Examples of Detailed Assessment Teams:

Metal finishing department in a large defense contractor:
- Metal finishing department manager
- Process engineer responsible for metal finishing processes
* Facilities engineer responsible for metal finishing department
- Wastewater treatment department supervisor
- Staff environmental engineer

Small pesticide formulator:
* Production supervisor
- Environmental engineer
- Maintenance engineer

Cyanide plating operation:
* Environmental engineer
- Electroplating facility engineering supervisor
- Plant chemist

Large offset printing facility:
Internal assessment team
    * Environmental engineer
    - Film processing supervisor
    - Pressroom supervisor
Outside assessment team (possible alternative team)
    * Engineer from within establishment
    - Environmental scientist
    - Printing industry technical consultant

---

* = Recommended team leader

Box 12

The box on the preceding page (Box 12) gives examples of assessment teams that might be designated for facilities of various sizes and in different industries. Note that for each team, the team leader is someone who has day-to-day operations responsibility and experience.

Review Data and Sites

Numerous data sources probably exist for a given site. Many of these may have been identified during the preliminary assessment. The detailed assessment team for that site will search for additional sources of data that will be useful in studying the targeted processes, operations, or waste streams.

Site reviews supplement and explain existing data.

However, most of their effort will be directed toward performing a thorough site review and interviewing workers. This will help them understand the data already collected and identify factors that are not well documented and for which data will need to be collected. Site review guidelines are outlined in Box 13.

Site reviews should be well planned.

- **Prepare an agenda** in advance that covers all points that still require clarification. Provide staff contacts in the area being assessed with the agenda several days before the inspection.
- **Schedule the inspection** to coincide with the particular operation that is of interest (e.g., makeup chemical addition, bath sampling, bath dumping, startup, shutdown, etc.).
- **Monitor the operation at different times** during all shifts, and if needed, during all three shifts, especially when waste generation is highly dependent on human involvement (e.g., in painting or parts cleaning operations).
- **Interview** the operators, shift supervisors, and work leaders in the assessed area. Discuss the waste generation aspects of the operation. Note their familiarity with the impacts their operation may have on other operations.
- **Photograph or videotape** the area of interest, if warranted. Pictures are valuable in the absence of plant layout drawings. Many details can be captured in pictures that otherwise could be forgotten or inaccurately recalled at a later date.
- **Observe the "housekeeping" aspects** of the operation. Check for signs of spills or leaks. Visit the maintenance shop and ask about problems in keeping the equipment leak-free. Assess the overall cleanliness of the site. Pay attention to odors and fumes.
- **Assess the organizational structure** and level of coordination of environmental activities between various departments.
- **Assess administrative controls**, such as cost accounting procedures, material purchasing procedures, and waste collection procedures.

Box 13

---

question and of how it fits into the overall facility operation. This perspective is a prerequisite for thorough assessment of options in later phases of the pollution prevention plan development cycle. If consultants are on the assessment team, the site review enables them to become familiar enough with the facility to utilize their expertise effectively.

The site review should not be performed perfunctorily, even though the assessment team members who are employed at the facility will all be familiar to some extent with the work-site being reviewed. Those who are not involved in the day-to-day operation in that area will see factors that otherwise would be overlooked. Furthermore, personnel assigned to that specific site will often see it in a new light when performing a pollution prevention assessment. Some of the information that can be gathered through site reviews is summarized in Box 14.

Typical questions to ask during site reviews include:

- What is the composition of the waste streams and emissions generated in the company? What is their quantity?
- From which production processes or treatments do these waste streams and emissions originate?
- Which waste materials and emissions fall under environmental regulations?
- What raw materials and input materials in the company or production process generate these waste streams and emissions?
- How much of a specific raw or input material is found in each waste stream?
- What quantity of materials are lost in the form of volatile emissions?
- How efficient is the production process and the various steps of that process?
- Are any unnecessary waste materials or emissions produced by mixing materials — which could otherwise be reused with other waste materials?
- Which good housekeeping practices are already in force in the company to limit the generation of waste materials?
- What process controls are already in use to improve process efficiency?

Box 14

Site visits should be well-planned to ensure that maximum benefit is obtained without excessive expenditures of time. While multiple visits to check or supplement data will usually be required, good planning can minimize such repetitions. Several suggestions for preparing for site visits are given below.

Good planning is essential for efficient site reviews.

Review existing documentation, such as operators' manuals and purchasing and shipping records. This will enable the team to focus on the topics to be investigated.

Decide on data sources and collection procedures.

Decide on data collection formats to ensure that the data collection will be rigorous and compatible with the compilation and analysis stage described on the following page. In particular, it is worthwhile to predetermine the boundaries and bases for calculating the energy and material balances that will be worked out during that stage. Doing a preliminary balance during the data collection phase can help identify data gaps and determine sampling requirements. The worksheets in Appendix A can be used for data collection, or you may decide to customize them or create entirely new ones to conform to the nature of the specific site. Appendices B and C may be helpful in developing new worksheets. **Photographs** are an excellent means of capturing extensive detail quickly and accurately.

Prepare an agenda and make sure that all team members and supervisors at the site receive it in advance.

Schedule site visits by contacting the staff in the area to be visited. Ask when they will be performing the operations you are particularly interested in assessing.

Observe operations as they are actually performed by different shifts and under various circumstances. Process units may be operated differently from the methods described in their operating manuals, or the equipment may have been modified without being so documented in the flow diagrams or equipment lists.

Look at procedures as they are performed in the production environment.

Interview workers and supervisors to determine how aware they are of what wastes are generated by their operation. They may have suggestions on reducing these wastes.

Follow the process from beginning to end, from the point where input materials enter the work-site to the point where products and wastes exit. This will help identify all suspected sources of waste. Waste sources to inspect include the production process; piping; maintenance operations; storage areas for raw materials, finished product, and work-in-process. Examine housekeeping practices and the waste treatment area, as well.

Identify waste sources.

Make follow-up visits as missing or unclear data are identified during the analysis stage.

Organize and Document Process Information

Analyzing process information involves preparing material and energy balances as a means of analyzing pollution sources and opportunities for eliminating them. Such a balance is an organized system of accounting for the flow, generation, consumption, and accumulation of mass and energy in a process. In its simplest form, a material balance is drawn up according to the mass conservation principle:

A material and energy balance for a given substance will reveal quantities lost to emission or to accumulation in equipment.

$$\text{Mass in} = \text{Mass out} - \text{Generation} + \text{Consumption} + \text{Accumulation}$$

If no chemical or nuclear reactions occur and the process progresses in a steady state, the material balance for any specific compound or constitutent is as follows:

$$\text{Mass out} = \text{Mass in}$$

The first step in preparing a balance is to draw a **process diagram**, which is a visual means of organizing the data on the energy and material flows and on the composition of the streams entering and leaving the system. Such a diagram shows the system boundaries, all streams entering and leaving the process, and points at which wastes are generated. An example of a flow diagram appears as Figure 4.

A process diagram organizes data graphically.

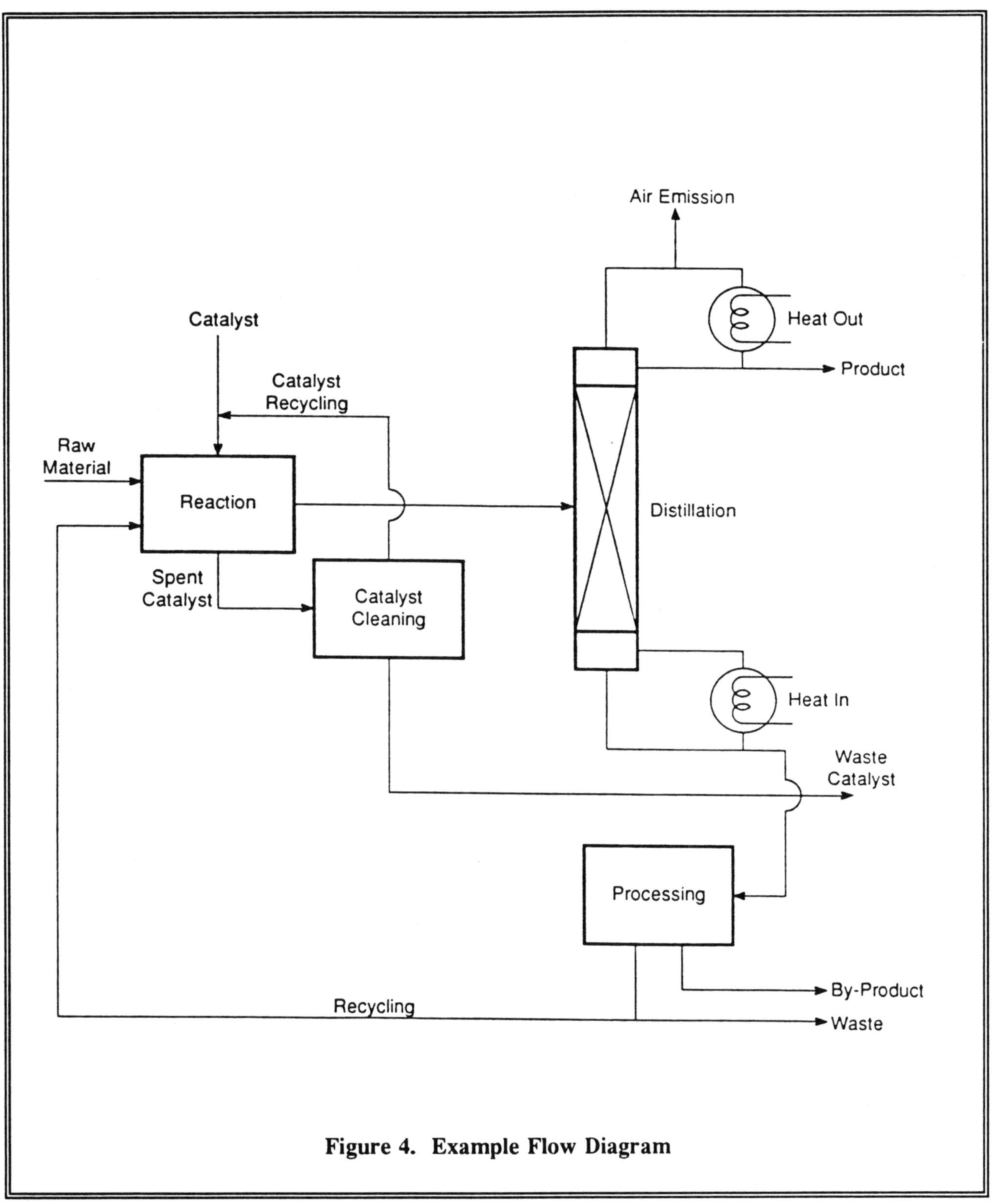

Figure 4. Example Flow Diagram

Boundaries should be selected according to the factors that are important for measuring the type and quantity of pollution prevented, the quality of the product, and the economics of the process. The amount of material input should equal the amount exiting, corrected for accumulation and creation or destruction.

A material balance should be calculated for each component entering and leaving the process. When chemical reactions take place in a system, there is an advantage to performing the material balance on the elements involved.

Each component should have a material balance calculated.

The limitations of material and energy balances should be understood. They are useful for organizing and extending pollution prevention data and should be used whenever possible. However, the user should recognize that most balance diagrams will be incomplete, approximate, or both.

Material and energy balances have some limitations.

- Most processes have numerous process streams, many of which affect various environmental media.
- The exact composition of many streams is unknown and cannot be easily analyzed.
- Phase changes occur within the process, requiring multimedia analysis and correlation.
- Plant operations or product mix change frequently, so the material and energy flows cannot be accurately characterized by a single balance diagram.
- Many sites lack sufficient historical data to characterize all streams.

These are examples of the complexities that will recur in analyzing real world processes.

Despite the limitations, material balances are essential to organize data, identify gaps, and permit estimation of missing information. They can help calculate concentrations of waste constituents where quantitative composition data are limited. They are particularly useful if there are points in the production process where it is difficult or uneconomical to collect or analyze samples. Data collection problems, such as an inaccurate reading or an unmeasured release, can be revealed when "mass in" fails to equal "mass out." Such an imbalance can also indicate that fugitive emissions are occurring. For example, solvent evaporation from a parts cleaning tank can be estimated as the difference between solvent put into the tank and solvent removed by disposal, recycling, or dragout.

Imbalances indicate that the data are inaccurate and should be reviewed or that fugitive emissions of waste are occurring.

DEFINE POLLUTION PREVENTION OPTIONS

Once the sources and nature of wastes generated have been described, the assessment team enters the creative phase. In a two-step procedure, they will propose and then screen pollution prevention options. Their objective is to generate a comprehensive set of options, ranked as to priority, that merit detailed feasibility assessment.

Propose Options

As with other planning efforts, the best results will be achieved in an environment that encourages creativity and independent thinking by each assessment team member. Brainstorming

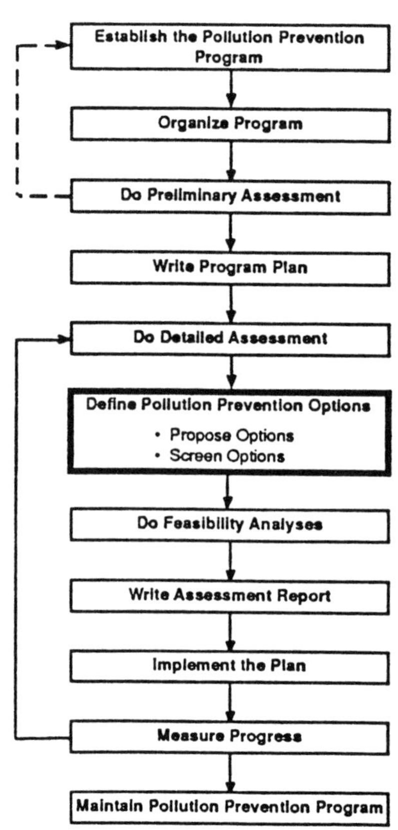

sessions are useful for encouraging creative thought because they provide a nonjudgmental, synergistic atmosphere in which ideas can be shared. Then, these ideas can be developed by means of group decision-making techniques.

This approach will enable the assessment team to identify options that the individual members might not have come up with on their own. Worksheet 7 in Appendix A is a suggested format for describing each option as it is proposed.

Structuring option definition sessions according to the U.S. EPA hierarchy (Chapter 1, Figure 1) will encourage the team to look first at true source reduction options, such as improved operating procedures and changes in technology, materials, and products. Then, options that involve reuse, or closed-loop recycling, would be examined. Finally, the team would consider off-line and off-site recycling and alternative treatment and disposal methods.

Screen Options

Many proposed options may result from the previous step. Since detailed technical, economic, and environmental feasibility analysis can be costly, the proposed options should be screened by the assessment team. Some options will be found to have no cost or risk attached; these can be implemented immediately. Others will be found to have marginal value or to be impractical; these will be dropped from further consideration. The remaining options will generally be found to require feasibility assessment.

This screening does not require detailed and costly study. Screening procedures can range from an informal review with a decision made by either the program manager or a vote of the team members, to the use of quantitative decision-making tools. Box 15 on the next page shows questions to be considered in option screening.

The informal review is a procedure by which the assessment team selects the options that appear best after discussing and examining each option. As is the case when the team is proposing options, their approach to screening should employ group decision-making techniques whenever possible.

In more complicated situations, the team may need to use the **weighted sum method** (see Appendix E) or another, similar technique designed for use in complex decision-making situations.

DO FEASIBILITY ANALYSES

The final product of the option definition phase is a prioritized list of pollution prevention options. These options now should be examined to determine which are technically, environmentally, and economically feasible and to prioritize them for implementation.

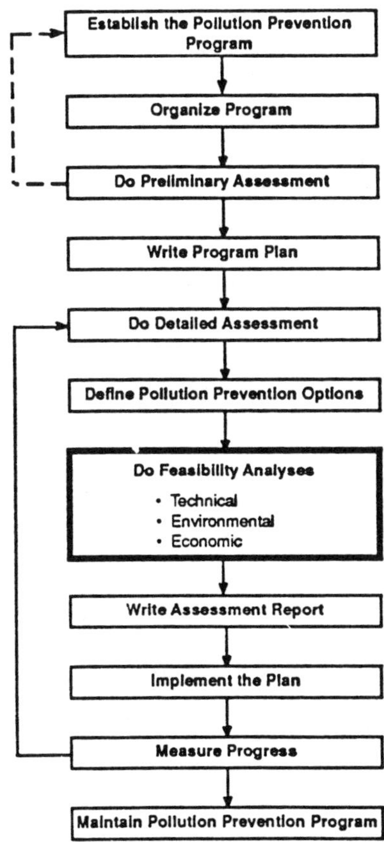

Option screening should consider these questions:

- Which options will best achieve the goal of waste reduction?
- What are the main benefits to be gained by implementing this option (e.g., financial, compliance, liability, workplace safety, etc.)?
- Does the necessary technology exist to develop the option?
- How much does it cost? Does it appear to be cost-effective, meriting in-depth economic feasibility assessment?
- Can the option be implemented within a reasonable amount of time without disrupting production?
- Does the option have a good "track record"? If not, is there convincing evidence that the option will work as required?
- What other areas will be affected?

Box 15

Depending on the resources currently available, it may be necessary to postpone feasibility assessments for some options. However, all options should be evaluated eventually.

Technical Evaluation

The assessment team will perform a technical evaluation to determine whether a proposed pollution prevention option is likely to work in a specific application. Technical evaluation for a given option may be relatively quick or it may require extensive investigation. The list in Box 16 suggests some criteria that could be used in a technical evaluation. Some of these are more detailed versions of questions asked during the option screening phase.

All groups in the facility that will be affected directly if the option is adopted should contribute to the technical evaluation. This might include people from production, maintenance, QC/QA, and purchasing. In some cases, customers may need to be consulted and their requirements verified. Prior consultation and review with these groups will ensure the viability and acceptance of an option. If the option calls for a change in production methods or input materials, carefully assess the likely effects on the quality of the final product. If after the technical evaluation the option appears impractical or can be expected to lower product quality, drop it.

Technical evaluations require the expertise of a variety of people.

For options that do not involve a significant capital expenditure, the team can use a "fast-track" approach. For example, procedural or housekeeping changes can often be implemented quickly, after the appropriate review, approvals, and training have been accomplished. Material substitutions also can be accomplished relatively quickly if there are no major production rate, product quality, or equipment changes involved.

Some options can be implemented right away.

Typical technical evaluation criteria:

- Will it reduce waste?
- Is the system safe for our workers?
- Will our product quality be improved or maintained?
- Do we have space available in our facility?
- Are the new equipment, materials, or procedures compatible with our production operating procedures, work flow, and production rates?
- Will we need to hire additional labor to implement the option?
- Will we need to train or hire personnel with special expertise to operate or maintain the new system?
- Do we have the utilities needed to run the equipment? Or, must they be installed at increased capital cost?
- How long will production be stopped during system installation?
- Will the vendor provide acceptable service?
- Will the system create other environmental problems?

Box 16

Equipment-related options or process changes are more expensive and may affect production rate or product quality. Therefore, such options require more study. The assessment team will want to determine whether the option will perform in the field under conditions similar to the planned application. In some cases, they can arrange, through equipment vendors and industry contacts, visits to existing installations. Experienced operators' comments are especially important and should be compared with vendors' claims. A bench-scale or pilot-scale demonstration may be needed. It may also be possible to obtain scale-up data using a rental test unit for bench-scale or pilot-scale experiments. Some vendors will install equipment on a trial basis, with acceptance and payment after a prescribed time, if the user is satisfied.

Options that can affect production or quality need careful study.

Environmental Evaluation

In this step, the pollution prevention assessment team will weigh the advantages and disadvantages of each option with regard to the environment. Often the environmental advantage is obvious — the toxicity of a waste stream will be reduced without generating a new waste stream. Most housekeeping and direct efficiency improvements have this advantage. With such options, the environmental situation in the company improves without new environmental problems arising.

Unfortunately, the environmental evaluation is not always so clearcut. Some options require a thorough environmental evaluation, especially if they involve product or process changes or the substitution of raw materials.

Environmental considerations:
- *effect on number and toxicity of waste streams*
- *risk of transfer to other media*
- *environmental impact of alternate input materials*
- *energy consumption*

For example, the engine rebuilding industry is dropping solvent and alkaline cleaners to remove grease and dirt from engines prior to disassembly. Instead, they are using high-temperature baking followed by shot blasting. This shift eliminates waste cleaner but presents a risk of atmospheric release because small quantities of components from the grease can vaporize.

To make a sound evaluation, the team should gather information on the environmental aspects of the relevant product, raw material or constituent part of the process. This information would consider the environmental effects not only of the production phase and product life cycle but also of extracting and transporting the alternative raw materials and of treating any unavoidable waste.

Energy consumption should also be considered. To make a sound choice, the evaluation should consider the entire life cycle of

Consider energy requirements.

both the product and the production process. Energy conservation is discussed in Chapter 8.

Economic Evaluation

Estimating the costs and benefits of some proposed pollution prevention projects is straightforward, while others prove to be complex. Despite the ease with which the cost calculations may be done for some options, it is advisable to document all that are adopted and to estimate the economic effects of each. This will help ensure that these real accomplishments of your pollution prevention program will not be overlooked when you measure the program's progress, as discussed in Chapter 4.

Document cost calculations so that the full benefit of the pollution prevention program can be quantified.

The economic feasibility needs to be checked and rechecked.

If a project has no significant capital costs, the decision is relatively simple. Its profitability can be judged by whether or not it reduces operating costs and/or prevents pollution. If it does, it can be implemented quickly. Installation of flow controls and improvement of operating practices, for example, probably will not require extensive analysis before they are adopted. Worksheet 9 (in Appendix A) can be used to document analysis of this type.

Operational changes usually can be installed quickly.

Projects with significant capital costs attached will require more detailed analysis. Worksheet 9 may be a good starting point, but an in-depth evaluation like the example that appears as Appendix F will be required.

There are a number of factors that make pollution prevention costs and benefits difficult to calculate for many proposed projects. The total costs of continuing to pollute are not discernible in most corporate accounting systems. Furthermore, many of these costs are probabilistic — although the risks are real, it is difficult to predict the cost and even the occurrence date from past experience. The long-term need to avoid the spiraling costs of waste treatment, storage, and disposal as well as future regulatory and liability entanglements are likely to be major elements of your pollution prevention project economic evaluation

Chapter 6 describes the Total Cost Analysis approach and gives an overview of the types of cost and benefit factors that should be examined when studying proposed pollution prevention projects. It suggests some approaches to calculating indirect and probabilistic costs so that their full impact can be included in economic feasibility assessments. It also discusses ways to track the economic effects of pollution prevention projects after they are implemented.

Most accounting systems do not reveal the total costs of continuing to pollute.

Total Cost Analysis is a useful mechanism for understanding the financial impact of pollution prevention projects.

WRITE THE ASSESSMENT REPORT

The task force should write a report that summarizes the results of the pollution prevention assessment at the company level. Box 17 shows the report contents. The report will provide a schedule for implementing prevention projects and will be the basis for evaluating and maintaining the pollution prevention program. It may also be needed to secure internal funding for projects that require capital investment, if the members of the pollution prevention assessment task force do not have the authority to commit funds.

You may be tempted to omit this step if your company has an owner-manager and only a few employees. A summary assessment report may not be needed to resolve pollution prevention project conflicts among different areas, and your funding approvals probably are not a formal procedure requiring cost justifications. However, an assessment report will help you focus subsequent pollution prevention efforts and will be useful as a record of what aspects of your business you examined for pollution prevention opportunities.

Input of the Assessment Teams

In a company that has several assessment teams, the task force will need to evaluate the results and resolve any conflicts that might exist among the teams as to approach and resources required for the projects they propose.

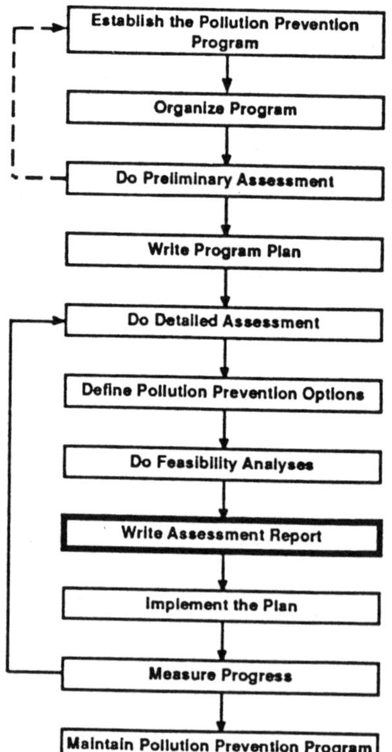

The report on each proposed project should discuss:

- Its pollution prevention potential
- The maturity of the technology and a discussion of successful applications
- The overall project economics
- The required resources and how they will be obtained
- The estimated time for installation and startup
- Possible performance measures to allow the project to be evaluated after it is implemented

Box 17

As input to this integration effort, each assessment team should prepare a summary report, presenting the results of their investigations and listing the options they screened. Each report should describe in some detail the options that the team has determined are feasible and propose a schedule for implementing them. The options recommended for immediate implementation should then be described in detail as proposed projects.

These proposals should evaluate each project under different scenarios. For example, the profitability of each could be estimated under both optimistic and pessimistic assumptions. Where appropriate, sensitivity analyses indicating the effect of key variables on profitability should be included. Each should outline a plan for adjusting and fine-tuning the initial projects as knowledge and experience increases. The proposals should include a schedule for addressing those areas and waste streams with lower priorities than the ones selected for the initial effort.

Preparing and Reviewing the Assessment Report

The task force will use the assessment teams' reports and project proposals to prepare the summary assessment report and implementation plan. The report should include a qualitative evaluation of the indirect and intangible costs and benefits to your company and employees of a pollution prevention plan. It will provide the basis for obtaining funding of pollution prevention projects. Pollution prevention projects should not be sold on their technical merits alone; a clear description of both tangible and intangible benefits can help a proposed project obtain funding.

Before the report is issued in final form, managers and other experienced people in the production units that will be affected by the proposed projects should be asked to review the report. Their review will help to ensure that the projects proposed are well-defined and feasible from their perspectives. While they probably were involved in the site reviews and other early efforts of the task force, they may spot inaccuracies or misunderstandings on the part of the assessment teams that were not apparent before.

In addition to ensuring the quality of the assessment report and implementation plan, this review will help ensure the support of the people who will be responsible for the success of the project.

IMPLEMENT THE POLLUTION PREVENTION PLAN

Select Projects for Implementation

Final decisions on which projects will be implemented and what the schedule will be are made at this point. If the task force or company executives question aspects of some projects, the assessment teams or pollution prevention program champions may be asked to produce additional data. They should be flexible enough to develop alternatives or modifications. They should also be willing to do background and support work, and they should anticipate potential problems in implementing the options. Above all, they should keep in mind that an idea will not sell if the marketers are not convinced.

Obtain Funding

The task force will seek to secure funding for those projects that will require expenditures. There will probably be other projects, such as expanding production capacity or moving into new product lines, that will compete with the pollution prevention program for funding. If the task force is part of the overall budget decision-making procedure, it can make an informed decision that a given pollution prevention project should be implemented right away or that it can wait until the next capital budgeting period. The task force will need to ensure that the project is reconsidered at that time.

Some companies will have difficulty raising funds internally for capital investment. If this applies to your company, look to outside financing. Private sector financing includes bank loans and other conventional sources of financing. Financial institutions are becoming more cognizant of the sound business aspects of pollution prevention.

Government financing is available in some cases. It may be worthwhile to contact your state's department of commerce or U.S. Small Business Administration for information regarding loans for pollution control. Some states can provide financial assistance. Appendix D includes a list of states providing this assistance and addresses where you can write for information.

Install the Selected Projects

Many pollution prevention projects will require changes in operating procedures, purchasing methods, or materials inventory control. Company policies and procedures documents and employee training will also be affected by the changes.

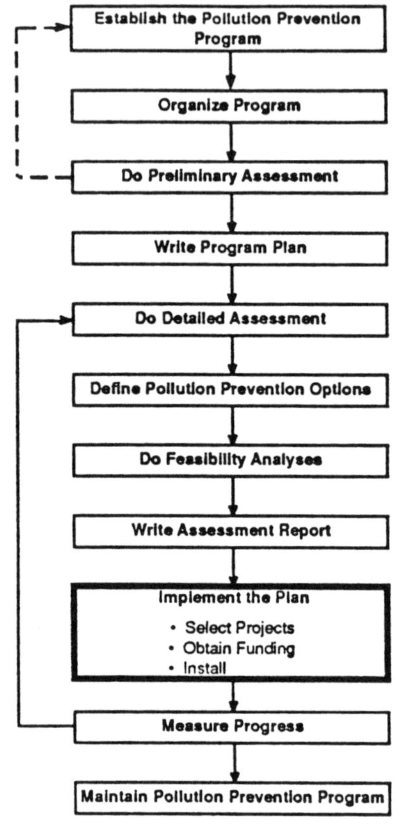

In 1989, the Bank of Boston started a unit focused strictly on environmental lending.

— Environmental Business Journal, October, 1991.

For projects that involve equipment modification or new equipment, the installation of a pollution prevention project is essentially the same as any other capital improvement project. The phases of the project include planning, design, procurement, construction, and operator training. As with other equipment acquisitions, it is important to get warranties from vendors prior to installation of the equipment.

Training and incentive programs may be needed to get employees used to the new pollution prevention procedures and equipment.

Review and Adjust

The pollution prevention process does not end with implementation. After the pollution prevention plan is implemented, track its effectiveness versus the claims made — technical, economic, etc. Options that do not meet your original performance expectations may require rework or modifications. Above all, reuse the knowledge gained by continuing to evaluate and fine-tune pollution prevention projects. Chapter 4 provides details on measuring progress after implementation and evaluating it against goals. Chapter 5 deals with ways to maintain and enhance a program after it is implemented.

You will want to measure your progress against your goals. By reviewing the program's successes and failures, managers at all levels can assess the degree to which pollution prevention goals at the facility and production unit levels are being met and what the economic results have been. The comparison identifies pollution prevention techniques that work well and those that do not. This information will help guide future pollution prevention assessment and implementation cycles.

Quantitative evaluation also enables you to compare your unit with similar units in your company and with data from other companies. You will need this knowledge to plan enhancements of your current pollution prevention program, to select technologies for transfer from other operations, and to help identify new pollution prevention options.

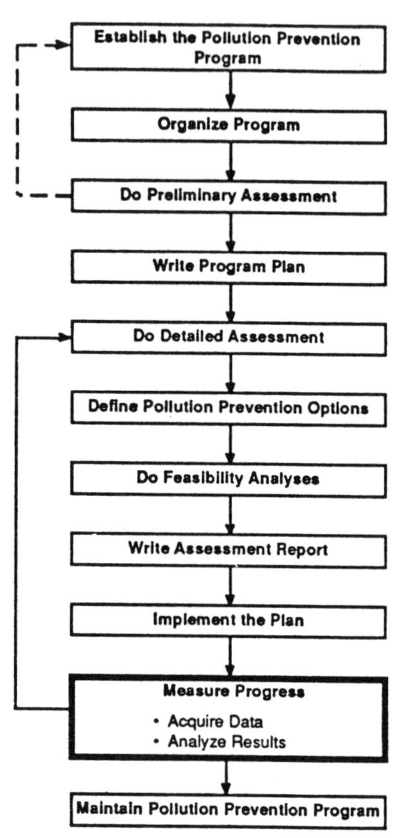

ACQUIRING DATA

You will need to select a quantity (e.g., waste volume or toxicity), measure that quantity, and normalize the data as necessary to correct for factors not related to the pollution prevention method being reviewed. Although the process is simple in theory, complexities arise in practice. There are a number of factors to consider when defining what data you will track.

First, the quantity selected to track performance must accurately reflect the waste(s) of interest. Second, the quantity must be measurable with the resources available to you. As in the Detailed Assessment Phase, material and energy balances will be helpful in organizing data and can help fill in some gaps in data.

After deciding what data should be tracked, you will need to determine how to collect it and what normalization may be required for each category of data.

Useful normalization factors include:

- *units produced*
- *hours of labor*
- *hours of production*

Regulatory Reporting Data

Depending on the type of business your company engages in, you may have a considerable volume of data already collected for regulated waste streams. However, there can be gaps and discrepancies in this data. For example, RCRA wastes are characterized by waste type and total amount, but not by individual components. Therefore, this data may not be specific enough for your evaluation. In addition, accurate measuring devices may not be available for all waste streams (e.g., vaporous or fugitive emissions). In

It may be necessary to supplement regulatory data.

such cases, your regulatory compliance reports would have been based on estimates; comparing estimates from one period to another will not yield very reliable percent-of-change figures. Finally, year-to-year comparisons may not be meaningful if the reporting requirements changed sufficiently to cause differences in how waste quantities were measured.

Wastes Shifted to Other Media

The pollution prevention option may eliminate part of the target material but shift some of it to another plant stream, to another environmental medium, or into the product.

It can be difficult to track the shift of a pollutant from one medium to another or to determine what new pollutants may be created by the new procedure. Transferring a given pollutant to another medium or replacing it with a different pollutant is, in principle, to be avoided. If you were to find that transfer was occurring, you would need to evaluate very carefully the relative impact on the environment.

Watch for shifts of wastes to other media.

Measuring Toxicity

The toxicity of the waste should be looked at, not just the quantity produced. Reducing the sheer volume of a given waste product while increasing its per-unit toxicity is a treatment option, but it is not pollution prevention. For example, adding lime to a waste stream to precipitate metals reduces the volume of waste but does not prevent pollution since the total quantity of metal is not changed. Since toxicity frequently is not measured as part of production reporting, you may have to establish procedures for doing so.

Toxicity measures may need to be developed.

Normalizing for External Factors

Changes in quantity are straightforward, easily understood, and relatively easy to calculate if data are available. Quantity comparisons from one period to the next can be useful input to a pollution prevention program review. However, the data will have to be normalized if there were major factors unrelated to pollution prevention efforts that influenced the quantities produced.

There are a number of external factors that can cause the quantities and/or mix of products and by-products to change. You will need to look carefully to see whether there are external factors for which you will need to normalize your data. Common ones to consider are: total hours the process operated; total employee hours; area, weight, or volume of product produced; number of batches processed; area, weight, or volume of raw material purchased; and profit from product. For continuous processes, the product output or raw material input can be a good normalization factor. Flow processes may be measured by volume or weight, whereas plating or film-making may be better normalized by area.

It may be necessary to normalize quantity comparisons to adjust for external factors.

In batch processes, production volume usually is related to waste production, but it may not be a linear relationship in all cases. For example, the quantity of solvent used at a printing plant is primarily a function of the total volume of stock printed and ink used, but it is also significantly influenced by the number of color changes made.

Another difficulty in comparing production and waste quantities arises when the relationship is inverse. This situation occurs frequently when the production rate decreases to the point that age-dated input materials in the inventory expire. For some production processes, waste is generated during start-up and shut-down of equipment. The volume of waste created in both situations is inversely proportional to the production volume.

Revenue and profit factors can indicate the amount of activity but may not be reliable indicators if market forces often cause prices to change. Thus, monetary factors typically apply only to products in stable markets.

Establishing a Baseline

When a pollution prevention option involves incremental changes to a well-defined process, it is possible to derive a baseline from historical performance. However, directly applicable historical data would not be available for new facilities.

Establishing a baseline is further complicated by changes to existing processes or equipment and by new facilities that are radically different from older plants for reasons other than pollution prevention alone. In this case, the measure of success may be the amount of pollution that was never generated. Thus, a projected amount of pollution may serve as a baseline.

METHODS OF ANALYZING THE DATA

As the above discussion indicates, measuring pollution prevention progress is complex. Therefore, using a single measure to summarize pollution prevention will be applicable only in the simplest cases, if at all. The characteristics of several approaches and their advantages and disadvantages are outlined in the following paragraphs. Select the method or combination of methods that best fits your data availability, facility characteristics, and corporate goals.

Semi-Quantitative Process Description

The semi-quantitative process description measurement method relies primarily on text, supplemented by a limited amount of numerical data. This type of analysis is less costly to prepare in terms of staff time and avoids many of the data collection problems discussed above. However, lack of quantitative data means that it has negligible value in evaluating achievement of specific

The...system monitors rates of toxic use and waste generation...avoiding distortion of production performance due to changes in overall volume of production.
— From an interview with Bill Schwalm on Polaroid's program, as reported in Environmental Business Journal, December, 1991.

Historical data may not be sufficient to establish baselines.

Select the most useful analysis method(s) for your situation.

Semi-quantitative methods are easier to prepare but have less utility.

goals. Lack of quantitative data also makes it difficult to compare similar processes when looking at potential technology transfer.

Quantity of Waste Shipped off Site or Treated on Site

Data for analysis based on transfers should be easy to obtain. Collecting such data for the SARA Title III chemicals is among the reporting requirements of the Pollution Prevention Act of 1990. Quantities of hazardous waste shipped off site are likely to be accurately recorded in manifests, although some inaccuracy may be introduced when partial barrels are shipped. In addition, the compositions of RCRA wastes may not be available in exact percentages. The amounts of trash and other nonhazardous wastes can be estimated based on shipment costs.

The amount of waste going to on-site waste treatment plants may be more difficult to obtain, but it should be possible to measure or estimate these quantities.

Shipping manifests and compliance reports provide data on quantities transferred.

Quantity of Materials Received

Changes in the quantities of materials brought on site, as determined from receiving records, can be used to measure pollution prevention progress. Most facilities keep detailed, accurate records of material received from suppliers. These records provide a source of data to track changes in the types and volumes of materials brought into the facility. However, this method may be difficult to apply at the process or project level. In addition, the quantity input will not accurately reflect the amount of waste if some of the material is destroyed during the process or is acquired from other production units in the facility,

Quantity purchased is an imprecise measure because it does not account for loss during processing.

Quantity of Waste Generated or Used

This method is a combination of the two previous ones. It essentially gives an overall material balance for each waste component. It involves tracking the quantities of hazardous, toxic, and other materials flowing into and out of the facility. It uses data on the quantities of material purchased, produced and destroyed in the production process, and incorporated in products and by-products, as well as discharges to waste treatment and disposal.

This approach gives an overall picture of material use but requires extensive data collection. Data on fugitive emissions are particularly difficult to track but can sometimes be estimated by calculating material balances.

Looking at both inputs and outputs provides a more complete understanding of progress.

Analysis of a Process

Pollution prevention can be measured on a process-by-process basis by examining the production process in detail for changes due to pollution prevention activities. If the process is carefully selected and can be defined precisely, this approach yields an

Analyses based strictly on processes will overlook facility-level waste, such as lighting and construction debris.

accurate description of process-related waste. It also allows better definition of a representative production or activity index for the waste generation.

However, it can be difficult to select which process to focus on in large facilities with complex, interconnected process units. The approach requires extensive data collection and analysis. In addition, many wastes may not be generated by a specific process. These nonprocess-specific wastes can be missed in a strictly process-based pollution prevention measurement system. Some types of waste that can be missed include construction debris, area lighting and utility support, and plant wastewater.

Analysis of a Pollution Prevention Project

This method focuses on measuring the results of each pollution prevention activity. It is suitable for facilities that produce many products from the same production line or for facilities that have a wide variety of production processes. As with the process approach, the data requirements are extensive. It also assumes a process orientation and thus is more applicable to product or equipment changes than to behavioral changes, such as good housekeeping or improved training.

Project analysis is more useful for production changes than for behavioral changes.

Change in Amount of Toxic Constituents

Pollution prevention can be measured by the change in the total amounts of toxic materials released. The data can be drawn in directly from SARA Title 313 Form R reporting. This method, obviously, does not apply to nonhazardous wastes.

Change in Material Toxicity

Testing for and eliminating the discharge of pollutants responsible for aquatic toxicity is required under the Clean Water Act. Pollutants causing aquatic toxicity may not be the pollutants on a "toxic pollutants list." The first class are compounds that are toxic to aquatic organisms and hence are assumed to be toxic to the aquatic environment. The second class are pollutants that have been tested on humans or other higher life forms and have been demonstrated to have detrimental effects.

Whole effluent toxicity testing is required under the NPDES permitting process. Standard methods are available to measure toxicity to aquatic life forms. The source of the toxicity can be identified by more detailed testing. Process streams contributing to the plant waste effluents are sampled and, if needed, partitioned into separate phases. This detailed toxicity testing allows identification and tracking of the actual toxicity of wastes from the plant. Toxicity testing requires sophisticated testing and data handling, however, and may not be feasible for all applications.

MEASURING ECONOMIC RESULTS

Aside from assessing its effectiveness in preventing pollution, a project should be evaluated like any other new process or capital investment. Preliminary cost estimates for installing and operating the system will be made prior to installing the system. More detailed data can be collected during construction and operation. The value of reduced waste production is estimated based on volumes of waste and cost of waste treatment and disposal. The economics of the process can then be evaluated by any of several techniques such as payback period, net present value, or return on investment.

Evaluate the cost effectiveness of the program.

The task of maintaining a viable pollution prevention program will made easier with the establishment of a **pollution prevention awareness program.** Such a program is intended to promote employee involvement in the prevention effort. The objectives of the pollution prevention awareness program are to:

- raise awareness of environment-related activities at the facility
- inform employees of specific environmental issues
- train employees in their pollution prevention responsibilities
- recognize employees for pollution prevention efforts
- encourage employees to participate in pollution prevention
- publicize success stories

A summary of methods for accomplishing this appears in Box 18 on the next page.

INTEGRATE POLLUTION PREVENTION INTO CORPORATE PLAN

Assign Accountability for Wastes

Operating units that generate wastes could be charged with the full costs of controlling and disposing of the wastes they generate. Cost accountability should also take into account indirect costs such as potential liability, compliance reporting, and oversight. Burying waste management costs in general overhead can lead to the illusion that disposal is "free." Allocating the costs of waste handling to the operating units that generate the waste reminds unit managers that waste control and disposal are increasingly large factors in the cost of doing business and motivates them to find ways to cease generating the waste. Chapter 6 describes several cost allocation methods.

Tracking and Reporting

Your information system should track and retain the data necessary to measure pollution prevention program results. You will need to ensure that these data are reviewed and reports prepared at meaningful intervals.

Pollution prevention is an ongoing effort that will be best maintained by personnel in the production area.

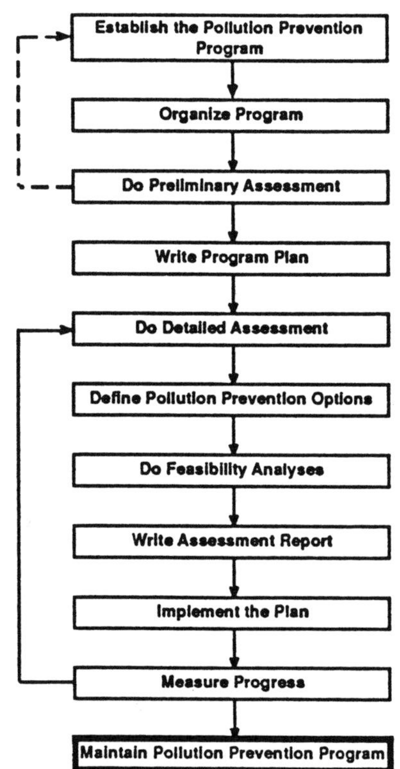

Key ways to maintain and improve the pollution prevention program:

- Integrate pollution prevention into corporate planning:
    - Assign pollution prevention accountability to the operating units where waste is generated
    - Track and report program status
    - Conduct an annual program evaluation at the corporate level

- Provide ongoing staff education programs:
    - Make pollution prevention awareness program a part of new employee orientation
    - Provide advanced training
    - Retrain supervisors and employees

- Maintain internal communication:
    - Encourage two-way communication between employees and management
    - Solicit employees' pollution prevention suggestions
    - Follow-up on suggestions

- Reward personnel for their success in pollution prevention:
    - Cite accomplishments in performance reviews
    - Recognize individual and group contributions
    - Grant material rewards
    - Consider pollution prevention a job responsibility subject to review

- Provide public outreach and education about pollution prevention efforts:.
    - Submit press releases on innovations to local media and to industry journals read by prospective clients.
    - Arrange for employees to speak publicly about pollution prevention measures in schools and civic organizations.

Box 18

Reports should be prepared frequently enough to enable unit managers to monitor and adjust their operations to adhere to the schedule that was established during the planning stage. (See Chapter 2.) In addition, they need this information to provide feedback to their staff, as discussed below.

Annual Program Evaluation

Top management can demonstrate continuing commitment to the program by conducting annual reviews of the program. The results of these annual reviews should be communicated to all employees through written announcements and meetings. Program successes should be recognized and any changes in objectives or policies announced and explained.

If these company-level reviews demonstrate chronic schedule
slippage, company executives and the pollution program task force
should meet to reevaluate the program. Some objectives or the
approach to achieving them may need to be adjusted. The purpose
is to maintain the same high profile the pollution prevention program had initially.

Tracking and reporting are essential.

STAFF EDUCATION

One of the most important elements of the waste minimization
and pollution prevention awareness program is training. The
training program should include all levels of personnel within the
company. The goal is to make each employee aware of waste
generation, its impact on the site and the environment, and ways
waste can be reduced and pollution prevented.

Classroom interaction generates ideas.

New Employee Orientation

A pollution prevention awareness orientation can be incorporated in the general orientation program given to all new employees. The orientation program would include the elements illustrated in Box 19.

Make sure new employees are aware of the program.

More detailed pollution prevention training should be provided to new employees after they have been on the job for a few weeks. This training will provide them with the skills they need to participate in pollution prevention. It also emphasizes the company's commitment to prevention.

At many plants, employees in certain jobs must be trained and examined on their knowledge of standard operating procedures specific to the site prior to working there. Pollution prevention training can be incorporated into this. It can also be incorporated in the QA procedures qualification process.

Make pollution prevention part of the QA process.

Advanced Training

Specialized training sessions on pollution prevention policy, procedures, and techniques should be provided to staff when their job scope is expanded or when they transfer to other areas in the company. These sessions should be considered part of the regular training program, and managers should have funds allocated to cover the costs.

Keep long-term employees' knowledge current.

If the progress of the pollution prevention program slows, review the amount and type of pollution prevention training provided and consider increasing its frequency and audience.

Example Pollution Prevention Employee Orientation

Course: "Pollution Prevention — Description, Motivation, and Practice"

Description: This training course emphasizes your company's commitment to pollution prevention. It gives instruction and practice in techniques for promotion, persuasion, and encouragement of pollution prevention.

Goal: The goal of the training program is to explain:
- What is pollution prevention?
- What leads to successful implementation of pollution prevention?
- What role can the individual play in promoting pollution prevention?

Lesson Plan for One-Day Orientation

Activities	Objectives
Get acquainted	Outline activities
Define terms and introduce objectives	Begin definition of pollution prevention as a concept and an activity
Group discussion	Perform and discuss a pollution prevention assessment of a simple process Outline pollution prevention opportunities Analyze implementation, possible barriers, and how to overcome
Hands-on exercise (1st half)	Perform and discuss pollution prevention assessment of a complex process
Form teams	Experience pressures of business
Individuals assigned roles	Experience importance of communication
Hands-on exercise (2nd half)	Refine application
Reassign roles	Develop teamwork
Repeat hands-on exercise (1st half)	Experience putting opportunities into priority list Discuss implementation, possible barriers, and how to overcome
Discussion	Reinforce need for pollution prevention Explain significance of individual contribution to pollution prevention

Box 19

Retraining

Periodic retraining of employees may be necessary when your policies and procedures change. Retraining employees also will reiterate your commitment to pollution prevention.

MAINTAIN INTERNAL COMMUNICATION

Two-Way Communication

Your goal is to keep employees motivated (see Box 20). They need to identify with and "buy in" to goals and objectives and continuously have the opportunity to contribute to its success. Employees will take their pollution prevention roles more seriously when their managers keep them informed and encourage them to submit pollution prevention ideas.

Make sure employees receive regularly scheduled status reports that are clear and truthful. Objectives that are described in vague terms and have poorly quantified results and reports that are issued at odd intervals may give the impression of a reduced priority for pollution prevention. Explain to the staff any schedule slippage resulting from unexpected challenges and the need for greater staff involvement, if applicable. Employees will work more effectively when they know what management expects of them. Cessation of reports or failure to show ongoing activities gives employees the impression that little progress is being made and/or that the overall program no longer is a priority.

Effective communication between managers and employees is a critical requirement for maintaining a successful program.

Solicit and Follow up on Employees' Suggestions

Employees' ideas for pollution prevention projects should be actively sought. Employees take their pollution prevention role more seriously when management keeps them informed and encourages them to submit pollution prevention ideas. Forums such as breakfasts or informal pollution prevention review meetings promote the exchange of information that will help generate new ideas. You could run a contest to get and reward employee input. For example, you could post a checklist of pollution prevention ideas and offer cash awards for the best way to implement an idea and for the best pollution prevention idea not included on the checklist.

Suggestions should be evaluated promptly and put into practice if they are found to be feasible. Similarly, if an employee submits an idea that is not implemented, explain why it was not used and work with the employee to develop a feasible idea. Prompt feedback is necessary to maintain employee interest.

Show employees their ideas are welcome.

To motivate employees, managers can:

- Provide feedback and reinforcement of employees' pollution prevention performance.

- Set an example by adhering to the pollution prevention program and actively considering employee ideas.

- Convey enthusiasm about meeting pollution prevention objectives.

- When new pollution prevention measures are implemented, explain how they fit in with the overall objectives.

- Regularly reinforce the importance of each individual's contributions to pollution prevention and their value to the overall objectives.

- Demonstrate personal commitment to the objectives and praise the commitment demonstrated by employees.

- Announce pollution prevention innovations by calling a meeting for all individuals who will be affected to discuss the change.

  - Open meeting to questions and comments.
  - Pay attention to signs of animosity or resistance and address these immediately.
  - Gain cooperation by showing that you know and care how the employees feel.

- Establish a "group identity" and work at building pride in adapting to the pollution prevention innovation.

- "Go to bat" for employees who have good pollution prevention ideas that have been rejected or overlooked.

- Establish quantifiable annual pollution prevention objectives:

  - On a monthly basis, have employees chart their personal and the company's progress against these objectives.
  - Incorporate pollution prevention goals, objectives, and accomplishments into annual job performance evaluations for people with direct process pollution prevention responsibilities.
  - Readjust objectives if they prove to be unattainable.

Box 20

EMPLOYEE REWARD PROGRAM

Performance Reviews

Progress in pollution prevention can be stated as an objective on which annual job performance evaluations are based, particularly at the management level. This delineates their responsibility for maintaining and enhancing the pollution prevention program. Using the formal mechanism of the written annual report to recognize efforts in this area raises the visibility of pollution prevention as something that is important to the company.

Recognition Among Peers

Employees who suggest pollution prevention measures that prove feasible and are slated for implementation should be publicized in the company newsletter or on bulletin boards. The estimated cost savings and/or other advantages that the company or unit will derive should be included in this announcement. Periodic group meetings may be a good forum for announcing individuals' efforts to control pollution in the company's daily operations.

Material Rewards

Cash or merchandise can be awarded to individuals. Establishing the award as a set percentage of the estimated annual savings to be realized by the company or production unit is one way to highlight the concrete value of pollution prevention.

Good suggestions should be put into practice and recognized.

PUBLIC OUTREACH AND EDUCATION

Employees can speak at meetings of community organizations and at schools to publicize the company's pollution prevention progress. Interviews with local media are another way to enhance corporate image and to further emphasize to employees the importance of the program.

Papers given at technical meetings and articles published in trade and professional journals are additional forms of positive publicity.

These measures all help to demonstrate that the company's commitment to pollution prevention is real.

Although businesses may invest in pollution prevention because it is the right thing to do or because it enhances their public image, the viability of many prevention investments rests on sound economic analyses. In essence, companies will not invest in a pollution prevention project unless that project successfully competes with alternative investments. The purpose of this chapter is to explain the basic elements of an adequate cost accounting system and how to conduct a comprehensive economic assessment of investment options.

A proposed pollution prevention option must compete with alternative investments.

TOTAL COST ASSESSMENT

In recent years industry and the EPA have begun to learn a great deal more about full evaluation of prevention-oriented investments. In the first place, we have learned that business accounting systems do not usually track environmental costs so they can be allocated to the particular production units that created those wastes. Without this sort of information, companies tend to lump environmental costs together in a single overhead account or simply add them to other budget line items where they cannot be disaggregated easily. As a result, companies do not have the ability to identify those parts of their operations that cause the greatest environmental expenditures or the products that are most responsible for waste production. This chapter provides some guidance on how accounting systems can be set up to capture this useful information better.

Standard accounting systems do not track environmental costs well.

It has also become apparent that economic assessments typically used for investment analysis may not be adequate for pollution prevention projects. For example, traditional analysis methods do not adequately address the fact that many pollution prevention measures will benefit a larger number of production areas than do most other kinds of capital investment. Second, they do not usually account for the full range of environmental expenses companies often incur. Third, they usually do not accommodate a sufficiently long time horizon to allow full evaluation of the benefits of many pollution prevention projects. Finally, they provide no mechanism for dealing with the probabilistic nature of pollution prevention benefits, many of which cannot be estimated with a high degree of certainty. This chapter provides guidance on how to overcome these problems as well.

Economic analysis of pollution prevention projects is complex because they:
- *affect multiple areas*
- *have long time horizons*
- *have probabilistic benefits*

In recognition of opportunities to accelerate pollution prevention, the U.S. EPA has funded several studies to demonstrate how economic assessments and accounting systems can be modified to improve the competitiveness of prevention-oriented investments. EPA calls this analysis **Total Cost Assessment** (TCA). There are four elements of Total Cost Assessment: expanded cost inventory, extended time horizon, use of long-term financial indicators, and direct allocation of costs to processes and products. The first three apply to feasibility assessment, while the fourth applies to cost accounting. Together these four elements will help you to demonstrate the true costs of pollution to your firm as well as the net benefits of prevention. In addition, they help you show how prevention-oriented investments compete with company-defined standards of profitability. In sum, TCA provides substantial benefits for pre-implementation feasibility assessments (see Chapter 2 on preliminary assessments and Chapter 3 on feasibility analysis) and for post-implementation project evaluation (see Chapter 4 on measuring progress.)

The remainder of this chapter summarizes the essential characteristics of TCA. Much of the information is drawn from a report recently prepared for the U.S. EPA by Tellus Institute. (See Appendix G for the full citation.) The Tellus report addresses TCA methodology in much greater detail than can be provided here and provides examples of specific applications from the pulp and paper industry. The report also includes an extensive bibliography on applying TCA to pollution prevention projects. In a separate but related study for the New Jersey Department of Environmental Protection, Tellus analyzed TCA as it applies to smaller and more varied industrial facilities. A copy of this report can be obtained from the N.J. Department of Environmental Protection.

EXPANDED COST INVENTORY

TCA includes not only the direct cost factors that are part of most project cost analyses but also indirect costs, many of which do not apply to other types of projects. Besides direct and indirect costs, TCA includes cost factors related to liability and to certain "less-tangible" benefits.

TCA is a flexible tool that can be adapted to your specific needs and circumstances. A full-blown TCA will make more sense for some businesses than for others. In either case it is important to remember that TCA can happen incrementally by gradually bringing each of its elements to the investment evaluation process. For example, while it may be quite easy to obtain information on direct costs, you may have more trouble estimating some of the future liabilities and less tangible costs. Perhaps your first effort should incorporate all direct costs and as many indirect costs as possible. Then you might add those costs that are more difficult to estimate as increments to the initial analysis, thereby highlighting to management both their uncertainty and their importance.

Direct Costs

For most capital investments, the direct cost factors are the only ones considered when project costs are being estimated. For pollution prevention projects, this category may be a net cost, even though a number of the components of the calculation will represent savings. Therefore, confining the cost analysis to direct costs may lead to the incorrect conclusion that pollution prevention is not a sound business investment.

Direct Costs
Capital Expenditures
- *Buildings*
- *Equipment and Installation*
- *Utility Connections*
- *Project Engineering*

Operation and Maintenance
Expenses or Revenues
- *Raw Materials*
- *Labor*
- *Waste Disposal*
- *Water and Energy*
- *Value of Recovered Material*

Indirect Costs

For pollution prevention projects, unlike more familiar capital investments, indirect costs are likely to represent a significant net savings. Administrative costs, regulatory compliance costs (such as permitting, recordkeeping, reporting, sampling, preparedness, closure/post-closure assurance), insurance costs, and on-site waste management and pollution control equipment operation costs can be significant. They are considered hidden in the sense that they are either allocated to overhead rather than their source (production process or product) or are altogether omitted from the project financial analysis. A necessary first step in including these costs in an economic analysis is to estimate and allocate them to their source. See the section below on Direct Cost Allocation for several ways to accomplish this.

Indirect Costs
Administrative Costs
Regulatory Compliance Costs
- *Permitting*
- *Recordkeeping and Reporting*
- *Monitoring*
- *Manifesting*

Insurance
Workman's Compensation
On-Site Waste Management
On-Site Pollution Control
  Equipment Operation

Liability Costs

Reduced liability associated with pollution prevention investments may also offer significant net savings to your company. Potential reductions in penalties, fines, cleanup costs, and personal injury and damage claims can make prevention investments more profitable, particularly in the long run.

In many instances, estimating and allocating future liability costs is subject to a high degree of uncertainty. It may, for example, be difficult to estimate liabilities from actions beyond your control, such as an accidental spill by a waste hauler. It may also be difficult to estimate future penalties and fines that might arise from noncompliance with regulatory standards that do not yet exist. Similarly, personal injury and property damage claims that may result from consumer misuse, from disposal of waste later classified as hazardous, or from claims of accidental release of hazardous waste after disposal are difficult to estimate. Allocation of future liabilities to the products or production processes also presents practical difficulties in a cost assessment. Uncertainty, therefore, is a significant aspect of a cost assessment and one that top management may be unaccustomed to or unwilling to accept.

Liability Costs
Penalties
Fines
Personal Injury
Property Damage
Natural Resources Damage Cleanup Costs
- *Superfund*
- *Corrective Action*

Some firms have nevertheless found alternative ways to address liability costs in project analysis. For example, in the narrative accompanying a profitability calculation, you could include a calculated estimate of liability reduction, cite a penalty or settlement that may be avoided (based on a claim against a similar company using a similar process), or qualitatively indicate without attaching dollar value the reduced liability risk associated with the pollution prevention project. Alternatively, some firms have chosen to loosen the financial performance requirements of their projects to account for liability reductions. For example, the required payback period can be lengthened from three to four years, or the required internal rate of return can be lowered from 15 to 10 percent. (See the U.S. EPA's *Pollution Prevention Benefits Manual, Phase II*, as referenced in Appendix G, for suggestions on formulas that may be useful for incorporating future liabilities into the cost analysis.)

Less-Tangible Benefits

A pollution prevention project may also deliver substantial benefits from an improved product and company image or from improved employee health. These benefits, listed in the cost allocation section of this chapter, remain largely unexamined in environmental investment decisions. Although they are often difficult to measure, they should be incorporated into the assessment whenever feasible. At the very least, they should be highlighted for managers after presenting the more easily quantifiable and allocatable costs.

Consider several examples. When a pollution prevention investment improves product performance to the point that the new product can be differentiated from its competition, market share may increase. Even conservative estimates of this increase can incrementally improve the payback from the pollution prevention investment. Companies similarly recognize that the development and marketing of so-called "green products" appeals to consumers and increasingly appeals to intermediate purchasers who are interested in incorporating "green" inputs into their products. Again, estimates of potential increases in sales can be added to the analysis. At the very least, the improved profitability from adding these less-tangible benefits to the analysis should be presented to management alongside the more easily estimated costs and benefits. Other less tangible benefits may be more difficult to quantify, but should nevertheless be brought to management's attention. For example, reduced health maintenance costs, avoided future regulatory costs, and improved relationships with regulators potentially affect the bottom line of the assessment.

In time, as the movement toward green products and companies grows, as workers come to expect safer working environments, and as companies move away from simply reacting to regulations and toward anticipating and addressing the environmental impacts of their processes and products, the less tangible

"We wanted to make a major effort to show that industry in the U.S. can simultaneously attack and solve environmental problems while improving both products and profitability."

— John Dudek, value analysis manager at Zytec, as quoted in Perspectives on Minnesota Waste Issues, January-February, 1992.

aspects of pollution prevention investments will become more apparent.

EXPANDED TIME HORIZON

Since many of the liability and less-tangible benefits of pollution prevention will occur over a long period of time, it is important that an economic assessment look at a long time frame, not the three to five years typically used for other types of projects. Of course, increasing the time frame increases the uncertainty of the cost factors used in the analysis.

Many of the benefits of pollution prevention accrue over long periods of time.

LONG-TERM FINANCIAL INDICATORS

When making pollution prevention decisions, select long-term financial indicators that account for:
* all cash flows during the project
* the time value of money.

Three commonly used financial indicators meet these criteria: Net Present Value (NPV) of an investment, Internal Rate of Return (IRR), and Profitability Index (PI). Another commonly used indicator, the Payback Period, does not meet the two criteria mentioned above and should not be used.

Discussions on using these and other indicators will be found in economic analysis texts.

Net Present Value, Internal Rate of Return, and Profitability Index are useful financial indicators.

DIRECT ALLOCATION OF COSTS

Few companies allocate environmental costs to the products and processes that produce these costs. Without direct allocation, businesses tend to lump these expenses into a single overhead account or simply add them to other budget line items where they cannot be disaggregated easily. The result is an accounting system that is incapable of (1) identifying the products or processes most responsible for environmental costs, (2) targeting prevention opportunity assessments and prevention investments to the high environmental cost products and processes, and (3) tracking the financial savings of a chosen prevention investment. TCA will help you remedy each of these deficiencies.

Like much of the TCA method, implementation of direct cost allocation should be flexible and tailored to the specific needs of your company. To help you evaluate the options available to you, the discussion below introduces three ways of thinking about allocating your costs: single pooling, multiple pooling, and service centers. The discussion is meant as general guidance and explains some of the advantages and disadvantages of each approach. Please see other EPA publications (such as those listed in Appen-

Developing a pollution prevention program may well provide the first real understanding of the costs of polluting.

Three methods of direct cost allocation:
* *single pooling*
* *multiple pooling*
* *service centers*

dix G), general accounting texts, and financial specialists for more detail.

Single Pool Concept

With the single pool method, the company distributes the benefits and costs of pollution prevention across all of its products or services. A general overhead or administrative cost is included in all transactions.

Advantages. This is the easiest accounting method to put into use. All pollution costs are included in the general or administrative overhead costs that most companies already have, even though they may not be itemized as pollution costs. It may therefore not be a change in accounting methods but rather an adjustment in the overhead rate. No detailed accounting or tracking of goods is needed. Little additional administrative burden is incurred to report the benefits of pollution prevention.

Disadvantages. If the company has a diverse product or service line, pollution costs may be recovered from products or services that do not contribute to the pollution. This has the effect of inflating the costs of those products or services unnecessarily. It also obscures the benefits of pollution prevention to the people who have the opportunity to make it successful — the line manager will not see the effect of preventing or failing to prevent pollution in his area of responsibility.

Single pool accounting is the easiest method, but it does not point up the effects of action within a given area.

Multiple Pool Concept

The next level of detail in the accounting process is the multiple pool concept, wherein pollution prevention benefits or costs are recovered at the department or other operating unit level.

Advantages. This approach ties the cost of pollution more closely to the responsible activity and to the people responsible for daily implementation. It is also easy to apply within an accounting system that is already set up for departmentalized accounting.

Disadvantages. A disparity may still exist between responsible activities and the cost of pollution. For example, consider a department that produces parts for many outside companies. Some customers need standard parts, while others require some special preparation of the parts. This special preparation produces pollution. Is it reasonable to allocate the benefit or cost for this pollution prevention project across all of the parts produced?

Multiple pool accounting comes closer to tracking responsibility.

Service Center Concept

A much more detailed level of accounting is the service center concept. Here, the benefits or costs of pollution prevention are allocated to only those activities that are directly responsible.

Advantages. Pollution costs are accurately tied to the generator. Theoretically, this is the most equitable to all products or services produced. Pollution costs can be identified as direct costs

Service center accounting applies costs or benefits to the activities that are directly responsible.

on the appropriate contracts and not buried in the indirect costs, affecting competitiveness on other contracts. Pollution costs are more accurately identified, monitored and managed. The direct benefits of pollution prevention are more easily identified and emphasized at the operational level.

Disadvantages. Considerable effort may be required to track each product, service, job, or contract and to recover the applicable pollution surcharges. Added administrative costs may be incurred to implement and maintain the system. It may be difficult to identify the costs of pollution when pricing an order or bidding on a new contract. It may be difficult to identify responsible activities under certain circumstances such as laboratory services where many small volumes of waste are generated on a seemingly continual basis.

SUMMARY

Environmental costs have been rising steadily for many years now. Initially, these costs did not seem to have a major impact on production. For this reason, most companies simply added these costs to an aggregate overhead account, if they tracked them at all. The tendency of companies to treat environmental costs as overhead and to ignore many of the direct, indirect, and less-tangible environmental costs (including future liability) in their investment decisions has driven the development of TCA.

Expanding your cost inventory pulls into your assessments a much wider array of environmental costs and benefits. Extending the time horizon, even slightly, can improve the profitability of prevention investments substantially, since these investments tend to have somewhat longer payback schedules. Choosing long-term financial indicators, which consistently provide managers with accurate and comparable project financial assessments, allows prevention oriented investments to compete successfully with other investment options. Finally, directly allocating costs to processes and products enhances your ability to target prevention investments to high environmental cost areas, routinely provides the information needed to do TCA analysis, and allows managers to track the success of prevention investments. Overall, the TCA method is a flexible tool, to be applied incrementally, as your company's needs dictate.

TCA is an increasingly valuable tool as the business costs of pollution continue to rise.

Environmentally compatible products minimize the adverse effects on the environment resulting from their manufacture, use, and disposal. The environmental impact of a product is to a large extent determined during its design phase. By taking environmental considerations into account during product planning, design, and development, your company can minimize the negative impact of your products on the environment.

What are environmentally compatible products?

Design changes made to prevent pollution should be implemented in such a manner that the quality or function of the product is not affected adversely. Design for the environment can be achieved by the people directly involved, within the framework of company policy and with support from company management, whether or not in response to incentives external to the company.

Compatibility can be integrated with other design concerns.

The process of looking at all aspects of product design from the preparation of its input materials to the end of its use is life-cycle assessment. A life-cycle assessment of the product design evaluates the types and quantities of product inputs, such as energy, raw materials, and water, and of product outputs, such as atmospheric emissions, solid and waterborne wastes, and the end-product.

STAGES IN LIFE-CYCLE ASSESSMENT

In 1990, the U.S. EPA sponsored an international pollution prevention conference on "clean" technologies and products. The introduction to the published proceedings (see Appendix G) provides the following overview.

Life-cycle assessment looks at all inputs and outputs of a product during its life cycle.

"Life-cycle assessment is a snapshot of inputs and outputs. It can be used as an objective technical tool to identify and evaluate opportunities to reduce the environmental impacts associated with a specific product, process, or activity. This tool can also be used to evaluate the effects of various resource management options designed to create sustainable systems. Life-cycle assessment takes a holistic approach by analyzing the entire life cycle ... encompassing extraction and processing (of) raw materials; manufacturing, transportation, and distribution; use/reuse/maintenance; recycling and composting; and final disposal.

"The three components of a life-cycle assessment include (1) the identification and quantification of energy and resource use and waste emissions (inventory analysis); (2) the assessment of the consequences those wastes have on the environment (impact analysis); and (3) the evaluation and implementation of opportunities to effect environmental improvements (improvement analysis). The life-cycle assessment is not necessarily a linear or stepwise process. Rather, information from any of the components can complement information from the other two. Environmental benefits can be realized from each component of the assessment process. For example, the **inventory** alone may be used to identify opportunities for reducing emissions, energy consumption, or material use. **Impact analysis** typically identifies the activities with greater and lesser environmental effects, while the **improvement analysis** helps ensure that any potential reduction strategies are optimized and that improvement programs do not produce additional, unanticipated adverse impacts to human health and the environment."

GOALS OF PRODUCT DESIGN OR REDESIGN

When beginning to look at product design or redesign to make it environmentally compatible, the first step is to define the goals. When redesigning an existing product, goals will involve modifying those aspects of its performance that are judged environmentally unacceptable and that can be improved. Aspects that should be examined include whether it uses a scarce input material, contains hazardous substances, uses too much energy, or is not readily reused or recycled. These environmental criteria can be added to the initial program of requirements for the product, such as quality, customer acceptance, and production price.

The goals of new product design can be reformulation and a rearrangement of the products' requirements to incorporate environmental considerations. For example, the new product can be made out of renewable resources, have an energy-efficient manufacturing process, have a long life, be non-toxic and be easy to reuse or recycle. In the design of a new product, these environmental considerations can become an integral part of the program of requirements.

J. C. van Weenan describes product design and redesign from the environmental impact perspective in his book *Waste Prevention: Theory and Practice*. (See Appendix G for the full reference.)

In both the redesign of existing products and the design of new products, the methods applied and the procedure followed will be affected by additional environmental requirements. These new environmental criteria will be added to the list of traditional criteria. Box 21 lists some environmental criteria for product design.

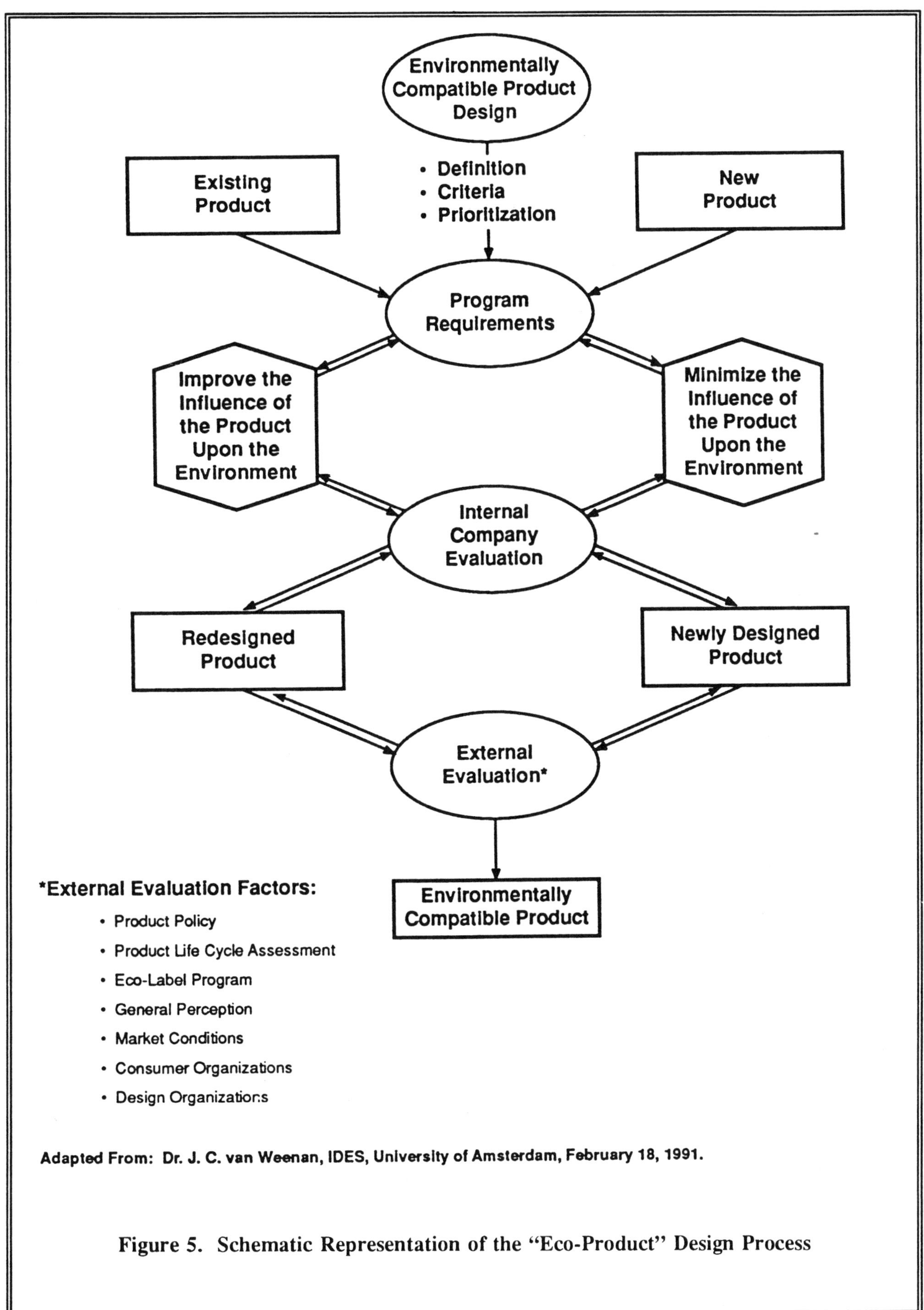

Figure 5. Schematic Representation of the "Eco-Product" Design Process

Environmental criteria to consider in designing products:

- Use renewable natural resource materials.
- Use recycled material.
- Use fewer toxic solvents or replace solvents with an alternative material (e.g., use bead blasting instead of solvents for paint removal).
- Reuse scrap and excess material.
- Use water-based inks instead of solvent-based ones.
- Produce combined or condensed products that reduce packaging requirements.
- Produce fewer integrated units (i.e., more replaceable component parts).
- Minimize product filler and packaging.
- Produce more durable products.
- Produce goods and packaging reusable by the consumer.
- Manufacture recyclable final products.

Box 21

The design process in Figure 5 shows a schematic representation of van Weenan's (1990) design of environmentally compatible products.

CHAPTER 8
ENERGY CONSERVATION
AND POLLUTION PREVENTION

Energy conservation and pollution prevention are complementary activities. That is, actions that conserve energy reduce the quantity of wastes produced by energy-generating processes, and actions that reduce production process wastes lower the expenditure of energy for waste handling and treatment.

Energy conservation goes hand-in-hand with pollution prevention.

PREVENTING POLLUTION BY CONSERVING ENERGY

Nearly all energy used in the United States is generated by processes that consume materials and create wastes that pollute the environment if released directly. These wastes require treatment or the even less satisfactory measure of long-term containment.

Wastes are produced in almost all energy-generating activities.

Wastes Produced by Energy Generation

Fossil fuel and nuclear power generation create a variety of wastes. The gaseous and particulate byproducts of fossil fuel combustion include carbon dioxide, carbon monoxide, and nitrogen and sulfur oxides. The processes used to treat these gases create other wastes. The use of nuclear energy presents the risk of accidental release of radioactive gases.

Water used in generating energy from fossil fuels is contaminated with the chemicals used to control scale and corrosion. Before discharge, the water must be treated to remove these contaminants. The water used in nuclear power plants can become contaminated accidentally, requiring that it be disposed of in a secure site.

Burning fossil fuels creates solid waste in the form of ash and slag. In addition, the treatment of waste gases and water causes the formation of solid waste. Waste nuclear fuel is another form of solid waste resulting from energy production.

Ways to Conserve Electrical and Thermal Energy

Production facilities consume enormous amounts of electricity in both their production processes and the operation of their facilities. Aside from environmental considerations, the rapid increase in the cost of electricity provides a strong motivation to conserve its use. Box 22 lists several ways to conserve electricity.

Consumption of electricity is a major cost for most facilities.

Your company can conserve electricity by:

- Implementing housekeeping measures such as turning off equipment and lights when not in use.
- Placing cool air intakes and air-conditioning units in cool, shaded locations.
- Using more efficient heating and refrigeration units.
- Using more efficient motors.
- Eliminating leaks in compressed air supply lines.
- Improving lubrication practices for motor-driven equipment.
- Using energy-efficient power transfer belts.
- Using fluorescent lights and/or lower wattage lamps or ballasts.
- Installing timers and/or thermostats to better control heating and cooling.

Box 22

Combustion of fossil fuels in primary heat sources such as boilers or fired heaters provides a major source of heat input to industrial processes. Thermal energy can be conserved by taking care to prevent its loss during transport from the combustion site to the specific processes where it is used. Box 23 lists some measures that can be taken to conserve thermal energy as it is transported and used. It may also be possible to recover and use heat generated by production processes.

You can reduce loss with thermal energy conservation by:

- Adjusting burners for optimal air/fuel ratio.
- Improving or increasing insulation on heating or cooling lines.
- Instituting regular maintenance to reduce leakage and stop steam trap bypass.
- Improving the thermodynamic efficiency of the process by options such as:

    — Using condensers or regenerative heat exchanger to recapture heat.
    — Using heat pumps or similar equipment to recover heat at distillation columns.
    — Using more efficient heat exchangers.
    — Using cogeneration of electricity and steam.

Box 23

CONSERVING ENERGY THROUGH POLLUTION PREVENTION

Energy consumption is reduced when waste generation is controlled. Treating and transporting pollutants represents an enormous drain on the energy reserves of the United States.

Pollution prevention activities result in improved efficiency of resource use, with a consequent reduction in the amount of energy required to process input materials. For example, reuse of metals such as copper or aluminum requires considerably less energy than is expended in extracting and processing the ores. Additional savings in energy can be realized by reducing the amount of metal used in a production process, thereby saving on energy required to recover the metal.

Two books listed in Appendix G deal specifically with facility energy conservation (Glasstone; Hu). They provide information on conducting energy audits, identifying conservation alternatives, and other topics related to improving the efficiency of energy use within a facility.

APPENDIX A
POLLUTION PREVENTION
WORKSHEETS

The worksheets in this appendix were designed to be useful at various points in the development of a pollution prevention program. Table A-1 lists the worksheets and describes the purpose of each.

Since these worksheets are intentionally generic, you may decide to redesign some or all of them to be more specific to your facility once you have your program underway. The checklists in Appendix B contain information that you may find helpful in deciding how to customize these worksheets to fit your situation. Appendix C contains examples of worksheets as they might be customized for a pharmaceutical company.

Table A-1. List of Pollution Prevention Assessment Worksheets

Phase	Number and Title	Purpose/Remarks
	1. Assessment Overview	Summarizes the overall program.
Assessment Phase		
	2. Site Description	Lists background information about the facility, including location, products, and operations.
	3. Process Information	This is a checklist of process information that can be collected before the assessment effort begins.
	4. Input Materials Summary	Records input material information for a specific production or process area. This includes name, supplier, hazardous component or properties, cost, delivery and shelf-life information, and possible substitutes.
	5. Products Summary	Identifies hazardous components, production rate, revenues, and other information about products.
	6. Waste Stream Summary	Summarizes the information collected for several waste streams. This sheet can be used to prioritize waste streams to assess.
	7. Option Generation	Records options proposed during brainstorming or nominal group technique sessions. Includes the rationale for proposing each option.
	8. Option Description	Describes and summarizes information about a proposed option. Also notes approval of promising options.
Feasibility Analysis Phase		
	9. Profitability	This worksheet is used to identify capital and operating costs and to calculate the payback period.

<table>
<tr>
<td>
Firm ___________________

Site ___________________

Date ___________________
</td>
<td>
**Pollution Prevention
Assessment Worksheets**

Proj. No. ___________________
</td>
<td>
Prepared By ___________________

Checked By ___________________

Sheet ___ of ___ Page ___ of ___
</td>
</tr>
</table>

WORKSHEET 1

ASSESSMENT OVERVIEW

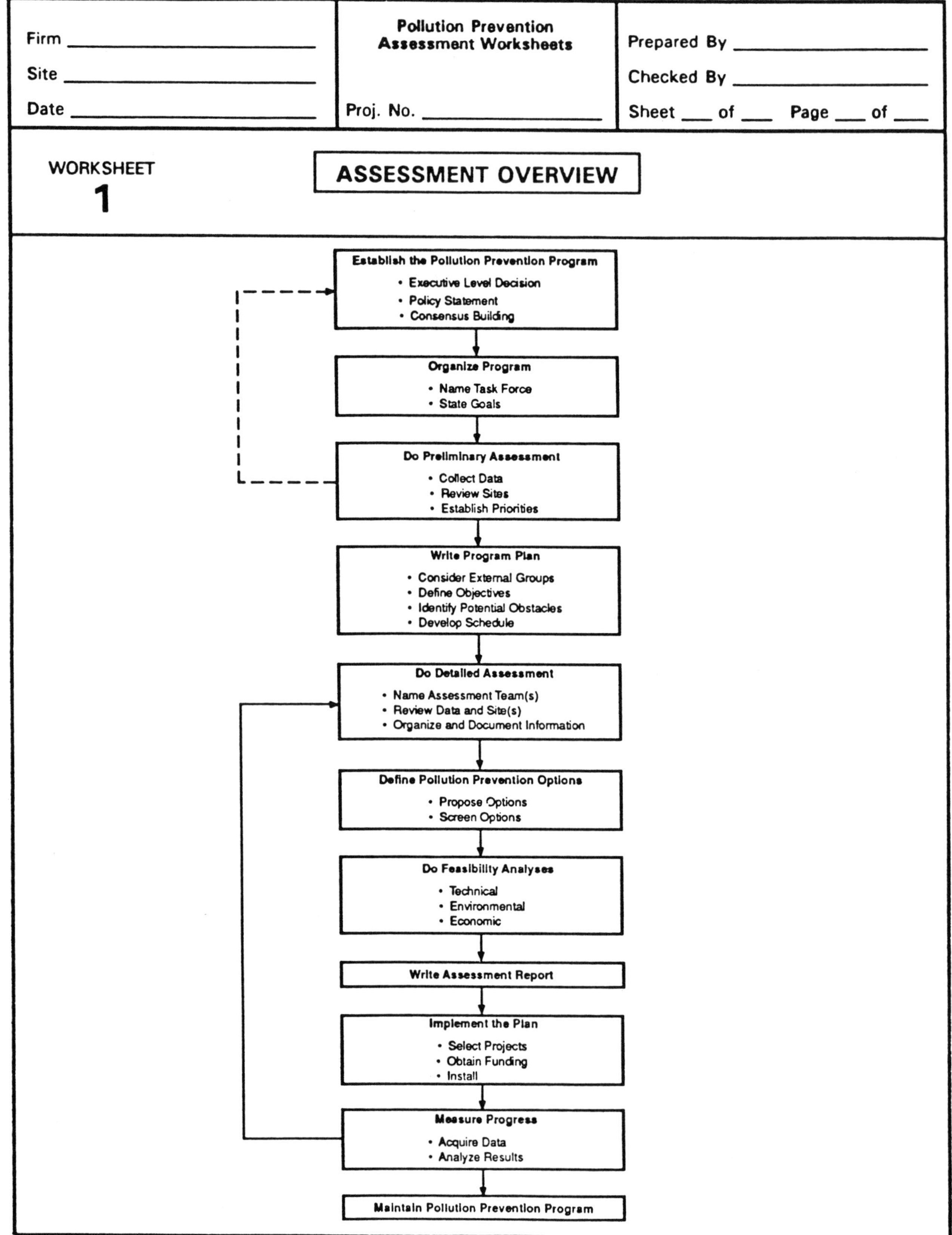

Appendix A

<table>
<tr><td>Firm ___________________
Site ___________________
Date ___________________</td><td>Pollution Prevention
Assessment Worksheets

Proj. No. _______________</td><td>Prepared By _______________
Checked By _______________
Sheet ___ of ___ Page ___ of ___</td></tr>
</table>

WORKSHEET 2 | **SITE DESCRIPTION**

Firm: ____

Plant: ____

Department: ____

Area: ____

Street Address: ____

City: ____

State/Zip Code: ____

Telephone: (   ) ____

Major Products: ____

____

SIC Codes: ____

EPA Generator Number: ____

Major Unit: ____

Product or Service: ____

Operations: ____

____

____

____

____

____

Facilities/Equipment Age: _______________________________________

____

____

____

<table>
<tr><td colspan="3">Firm _______________

Site _______________

Date _______________</td><td colspan="3">Pollution Prevention
Assessment Worksheets

Proj. No. _______________</td><td colspan="3">Prepared By _______________

Checked By _______________

Sheet ___ of ___   Page ___ of ___</td></tr>
</table>

WORKSHEET	**PROCESS INFORMATION**
3	

Process Unit/Operation: _______________________________

Operation Type: ☐ Continuous   ☐ Discrete

               ☐ Batch or Semi-Batch   ☐ Other ___________

Document	Complete? (Y/N)	Current? (Y/N)	Last Revision	Used in this Report (Y/N)	Document Number	Location
Process Flow Diagram						
Material/Energy Balance						
Design						
Operating						
Flow/Amount Measurements						
Stream						
Analyses/Assays						
Stream						
Process Description						
Operating Manuals						
Equipment List						
Equipment Specifications						
Piping and Instrument Diagrams						
Plot and Elevation Plan(s)						
Work Flow Diagrams						
Hazardous Waste Manifests						
Emission Inventories						
Annual/Biennial Reports						
Environmental Audit Reports						
Permit/Permit Applications						
Batch Sheet(s)						
Materials Application Diagrams						
Product Composition Sheets						
Material Safety Data Sheets						
Inventory Records						
Operator Logs						
Production Schedules						

        Appendix A

<table>
<tr><td>Firm _______________
Site _______________
Date _______________</td><td>**Pollution Prevention
Assessment Worksheets**

Proj. No. _______________</td><td>Prepared By _______________
Checked By _______________
Sheet ___ of ___ Page ___ of ___</td></tr>
</table>

WORKSHEET
4

INPUT MATERIALS SUMMARY

Attribute	Description		
	Stream No. ____	Stream No. ____	Stream No. ____
Name/ID			
Source/Supplier			
Component/Attribute of Concern			
Annual Consumption Rate			
Overall			
Component(s) of Concern			
Purchase Price, $ per ________			
Overall Annual Cost			
Delivery Mode[1]			
Shipping Container Size & Type[2]			
Storage Mode[3]			
Transfer Mode[4]			
Empty Container Disposal Management[5]			
Shelf Life			
Supplier Would			
— accept expired material? (Y/N)			
— accept shipping containers? (Y/N)			
— revise expiration date? (Y/N)			
Acceptable Substitute(s), if any			
Alternate Supplier(s)			

Notes:
1. e.g., pipeline, tank car, 100 bbl tank truck, truck, etc.
2. e.g., 55 gal drum 100 lb paper bag, tank, etc.
3. e.g., outdoor, warehouse, underground, aboveground, etc.
4. e.g., pump, forklift, pneumatic transport, conveyor, etc.
5. e.g., crush and landfill, clean and recycle, return to supplier, etc.

Firm ___________________________	**Pollution Prevention** **Assessment Worksheets**	Prepared By _________________
Site ___________________________		Checked By _________________
Date ___________________________	Proj. No. ________________	Sheet ___ of ___ Page ___ of ___

<table>
<tr><td colspan="2">WORKSHEET
5</td><td colspan="3">**PRODUCTS SUMMARY**</td></tr>
<tr><td rowspan="2">Attribute</td><td colspan="3">Description</td></tr>
<tr><td>Stream No. _____</td><td>Stream No. _____</td><td>Stream No. _____</td></tr>
<tr><td>Name/ID</td><td></td><td></td><td></td></tr>
<tr><td></td><td></td><td></td><td></td></tr>
<tr><td>Component/Attribute of Concern</td><td></td><td></td><td></td></tr>
<tr><td></td><td></td><td></td><td></td></tr>
<tr><td>Annual Production Rate</td><td></td><td></td><td></td></tr>
<tr><td> Overall</td><td></td><td></td><td></td></tr>
<tr><td> Component(s) of Concern</td><td></td><td></td><td></td></tr>
<tr><td></td><td></td><td></td><td></td></tr>
<tr><td>Annual Revenues, $ _________</td><td></td><td></td><td></td></tr>
<tr><td></td><td></td><td></td><td></td></tr>
<tr><td>Shipping Mode</td><td></td><td></td><td></td></tr>
<tr><td>Shipping Container Size & Type</td><td></td><td></td><td></td></tr>
<tr><td>Onsite Storage Mode</td><td></td><td></td><td></td></tr>
<tr><td>Containers Returnable (Y/N)</td><td></td><td></td><td></td></tr>
<tr><td>Shelf Life</td><td></td><td></td><td></td></tr>
<tr><td>Rework Possible (Y/N)</td><td></td><td></td><td></td></tr>
<tr><td>Customer Would</td><td></td><td></td><td></td></tr>
<tr><td> — relax specification (Y/N)</td><td></td><td></td><td></td></tr>
<tr><td> — accept larger containers (Y/N)</td><td></td><td></td><td></td></tr>
<tr><td></td><td></td><td></td><td></td></tr>
<tr><td></td><td></td><td></td><td></td></tr>
<tr><td></td><td></td><td></td><td></td></tr>
<tr><td></td><td></td><td></td><td></td></tr>
<tr><td></td><td></td><td></td><td></td></tr>
<tr><td></td><td></td><td></td><td></td></tr>
</table>

<table>
<tr><td colspan="2">Firm _______________________

Site _______________________

Date _______________________</td><td colspan="2">Pollution Prevention
Assessment Worksheets

Proj. No. _______________</td><td colspan="2">Prepared By _______________

Checked By _______________

Sheet ___ of ___   Page ___ of ___</td></tr>
</table>

WORKSHEET 6

WASTE STREAM SUMMARY

Attribute		Description					
		Stream No. _____		Stream No. _____		Stream No. _____	
Waste ID/Name:							
Source/Origin							
Component or Property of Concern							
Annual Generation Rate (units ________)							
Overall							
Component(s) of Concern							
Cost of Disposal							
Unit Cost ($ per: ________)							
Overall (per year)							
Method of Management[1]							

Priority Rating Criteria[2]	Relative Wt. (W)	Rating (R)	R x W	Rating (R)	R x W	Rating (R)	R x W
Regulatory Compliance							
Treatment/Disposal Cost							
Potential Liability							
Waste Quantity Generated							
Waste Hazard							
Safety Hazard							
Minimization Potential							
Potential to Remove Bottleneck							
Potential By-product Recovery							
Sum of Priority Rating Scores		Σ(RxW)		Σ(RxW)		Σ(RxW)	
Priority Rank							

Notes:
1. For example, sanitary landfill, hazardous waste landfill, on-site recycle, incineration, combustion with heat recovery, distillation, dewatering, etc.

2. Rate each stream in each category on a scale from 0 (none) to 10 (high).

<table>
<tr><td>Firm _______________
Site _______________
Date _______________</td><td>**Pollution Prevention
Assessment Worksheets**

Proj. No. _______________</td><td>Prepared By _______________
Checked By _______________
Sheet ___ of ___ Page ___ of ___</td></tr>
</table>

WORKSHEET 7

OPTION GENERATION

Meeting format (e.g., brainstorming, nominal group technique) _______________________________

Meeting Coordinator ____

Meeting Participants ____

____

____

____

List Suggestion Options	Rationale/Remarks on Option

| Firm _______________________

Site _______________________

Date _______________________ | **Pollution Prevention**
Assessment Worksheets

Proj. No. _______________________ | Prepared By _______________________

Checked By _______________________

Sheet ___ of ___ Page ___ of ___ |

WORKSHEET

8

OPTION DESCRIPTION

Option Name: ____

Briefly describe the option: ____

____

____

____

____

____

____

____

____

Waste Stream(s) Affected: ____

____

Input Material(s) Affected: ____

____

Product(s) Affected: ____

____

Indicate Type:

☐ Source Reduction
- _______ Equipment-Related Change
- _______ Personnel/Procedure-Related Change
- _______ Materials-Related Change

☐ Recycling/Reuse
- _______ Onsite   _______ Material reused for original purpose
- _______ Offsite _______ Material used for a lower-quality purpose
- _______ Material sold

Originally proposed by: _________________________________   Date: ___________

Reviewed by: ____   Date: ___________

Approved for study? __________ yes __________ no   By:______________________

Reason for Acceptance or Rejection ____

Firm _________________________ Site _________________________ Date _________________________	**Pollution Prevention Assessment Worksheets** Proj. No. _________________	Prepared By _________________ Checked By _________________ Sheet ___ of ___ Page ___ of ___

WORKSHEET # 9	**PROFITABILITY**

Capital Costs

Purchased Equipment ____

Materials ____

Installation ____

Utility Connections ____

Engineering ____

Start-up and Training ____

Other Capital Costs ____

Total Capital Costs ____

Incremental Annual Operating Costs

Change in Disposal Costs ____

Change in Raw Material Costs ____

Change in Other Costs ____

Annual Net Operating Cost Savings ____

$$\text{Payback Period (in years)} = \frac{\text{Total Capital Costs}}{\text{Annual Net Operating Cost Savings}} = \underline{\hspace{6cm}}$$

APPENDIX B
INDUSTRY-SPECIFIC CHECKLISTS

This appendix tabulates information that may be helpful to you if you decide to customize the worksheets in Appendix A for your own company's needs. Some ideas for achieving pollution prevention through good operating practices are shown in Table 1. Approaches to pollution prevention in material receiving, raw material and product storage, laboratories, and maintenance areas are shown in Table 2. Information in these two tables can apply to a wide range of industries. Industry-specific checklists for five example industries are presented in Tables 3 through 7. See Appendix G for a list of publications that provide industry-specific information related to pollution prevention. The tables contained within this appendix are as follows:

Table 1. Pollution Prevention Through Good Operating Practices
Table 2. Checklist for All Industries
Table 3. Checklist for the Printing Industry
Table 4. Checklist for the Fabricated Metal Industry
Table 5. Checklist for the Metal Casting Industry
Table 6. Checklist for the Printed Circuit Board Industry
Table 7. Checklist for the Coating Industry

Table 1. Pollution Prevention Through Good Operating Practices

Good Operating Practice	Program Ingredients
Waste Segregation	Prevent mixing of hazardous wastes with nonhazardous wastes Store materials in compatible groups Segregate different solvents Isolate liquid wastes from solid wastes
Preventive Maintenance Programs	Maintain equipment history cards on equipment location, characteristics, and maintenance Maintain a master preventive maintenance (PM) schedule Keep vendor maintenance manuals handy Maintain a manual or computerized repair history file
Training/Awareness-Building Programs	Provide training for - Operation of the equipment to minimize energy use and material waste - Proper materials handling to reduce waste and spills - Emphasize importance of pollution prevention by explaining the economic and environmental ramifications of hazardous waste generation and disposal - Detecting and minimizing material loss to air, land, or water - Emergency procedures to minimize lost materials during accidents
Effective Supervision	Closer supervision may improve production efficiency and reduce inadvertent waste generation Centralize waste management. Appoint a safety/waste management officer for each department. Educate staff on the benefits of pollution prevention. Establish pollution prevention goals. Perform pollution prevention assessments.
Employee Participation	"Quality circles" (free forums between employees and supervisors) can identify ways to reduce waste Solicit and reward employee suggestions for waste reduction ideas

Good Operating Practice	Program Ingredients
Production Scheduling/Planning	Maximize batch size to reduce clean out waste
	Dedicate equipment to a single product
	Alter batch sequencing to minimize cleaning frequency (light-to-dark batch sequence, for example)
Cost accounting/ Allocation	Charge direct and indirect costs of all air, land, and water discharges to specific processes or products
	Allocate waste treatment and disposal costs to the operations that generate the waste
	Allocate utility costs to specific processes or products

Table 2. Checklist for All Industries

Waste Origin/Type	Pollution Prevention and Recycling Methods
Material Receiving/ Packaging materials, off-spec materials, damaged container, inadvertent spills, transfer hose emptying	Use "Just-in-Time" ordering system. Establish a centralized purchasing program. Select quantity and package type to minimize packing waste. Order reagent chemicals in exact amounts. Encourage chemical suppliers to become responsible partners (e.g., accept outdated supplies). Establish an inventory control program to trace chemical from cradle to grave. Rotate chemical stock. Develop a running inventory of unused chemicals for other departments' use. Inspect material before accepting a shipment. Review material procurement specifications. Validate shelf-life expiration dates. Test effectiveness of outdated material. Eliminate shelf-life requirements for stable compounds. Conduct frequent inventory checks. Use computer-assisted plant inventory system. Conduct periodic materials tracking. Properly label all containers. Set up staffed control points to dispense chemicals and collect wastes. Buy pure feeds. Find less critical uses for off-spec material (that would otherwise be disposed). Change to reusable shipping containers. Switch to less hazardous raw material. Use rinsable/recyclable drums.
Raw Material and Product Storage/ Tank bottoms; off-spec and excess materials; spill residues; leaking pumps, valves, tanks, and pipes; damaged containers; empty containers	Establish Spill Prevention, Control, and Countermeasures (SPCC) plans. Use properly designed tanks and vessels only for their intended purposes. Install overflow alarms for all tanks and vessels. Maintain physical integrity of all tanks and vessels. Set up written procedures for all loading/unloading and transfer operations. Install secondary containment areas. Instruct operators to not bypass interlocks, alarms, or significantly alter setpoints without authorization. Isolate equipment or process lines that leak or are not in service. Use sealless pumps.

Waste Origin/Type	Pollution Prevention and Recycling Methods
Raw Material and Product Storage/ (Continued)	Use bellows-seal valves.
	Document all spillage.
	Perform overall materials balances and estimate the quantity and dollar value of all losses.
	Use floating-roof tanks for VOC control.
	Use conservation vents on fixed roof tanks.
	Use vapor recovery systems.
	Store containers in such a way as to allow for visual inspection for corrosion and leaks.
	Stack containers in a way to minimize the chance of tipping, puncturing, or breaking.
	Prevent concrete "sweating" by raising the drum off storage pads.
	Maintain Material Safety Data Sheets to ensure correct handling of spills.
	Provide adequate lighting in the storage area.
	Maintain a clean, even surface in transportation areas.
	Keep aisles clear of obstruction.
	Maintain distance between incompatible chemicals.
	Maintain distance between different types of chemicals to prevent cross-contamination.
	Avoid stacking containers against process equipment.
	Follow manufacturers' suggestions on the storage and handling of all raw materials.
	Use proper insulation of electric circuitry and inspect regularly for corrosion and potential sparking.
	Use large containers for bulk storage whenever possible.
	Use containers with height-to-diameter ratio equal to one to minimize wetted area.
	Empty drums and containers thoroughly before cleaning or disposal.
	Reuse scrap paper for note pads; recycle paper.
Laboratories/ Reagents, off-spec chemicals, samples, empty sample and chemical containers	Use micro or semi-micro analytical techniques.
	Increase use of instrumentation.
	Reduce or eliminate the use of highly toxic chemicals in laboratory experiments.
	Reuse/recycle spent solvents.
	Recover metal from catalyst.

Table 2. (Continued)

Waste Origin/Type	Pollution Prevention and Recycling Methods
Laboratories (Continued)	Treat or destroy hazardous waste products as the last step in experiments.
	Keep individual hazardous waste streams segregated, segregate hazardous waste from nonhazardous waste, segregate recyclable waste from non-recyclable waste.
	Assure that the identity of all chemicals and wastes is clearly marked on all containers.
	Investigate mercury recovery and recycling.
Operation and Process Changes Solvents, cleaning agents, degreasing sludges, sandblasting waste, caustic, scrap metal, oils, greases from equipment cleaning	Maximize dedication of process equipment.
	Use squeegees to recover residual fluid on product prior to rinsing.
	Use closed storage and transfer systems.
	Provide sufficient drain time for liquids.
	Line equipment to reduce fluid holdup.
	Use cleaning system that avoid or minimize solvents and clean only when needed.
	Use countercurrent rinsing.
	Use clean-in-place systems.
	Clean equipment immediately after use.
	Reuse cleanup solvent.
	Reprocess cleanup solvent into useful products.
	Segregate wastes by solvent type.
	Standardize solvent usage.
	Reclaim solvent by distillation.
	Schedule production to lower cleaning frequency.
	Use mechanical wipers on mixing tanks.
Operation and Process Changes Sludge and spent acid from heat exchanger cleaning	Use bypass control or pumped recycle to maintain turbulence during turndown.
	Use smooth heat exchange surfaces.
	Use on-stream cleaning techniques.
	Use high pressure water cleaning to replace chemical cleaning where possible.
	Use lower pressure steam.

Table 3. Checklist for the Printing Industry

Waste Origin/Type	Pollution Prevention and Recycling Method
Image Processing/Empty containers, used film packages, outdated material	Recycle empty containers. Recycle spoiled photographic film.
Image Processing/ Photographic chemicals, silver	Use silver-free films, such as vesicular, diazo, or electrostatic types.. Use water-developed litho plates. Extend bath life. Use squeegees to reduce carryover. Employ countercurrent washing. Recover silver and recycle chemicals.
Plate Making/Damaged plates, developed film, outdated materials	Use electronic imaging, laser plate making.
Plate Making/ Acids, alkali, solvents, plate coatings (may contain dyes, photopolymers, binders, resins, pigment, organic acids), developers (may contain isopropanol, gum arabic, lacquers, caustics), and rinse water	Electronic imaging/laser print making. Recover silver and recycle chemicals. Use floating lids on bleach and developer tanks. Use countercurrent washing sequence. Use squeegees to reduce carryover. Substitute iron-EDTA for ferrocyanide. Use washless processing systems. Use better operating practices. Remove heavy metals from wastewater.
Finishing/Damaged products, scrap	Reduce paper use and recycle waste paper.
Printing/ Lubricating oils, waste ink, cleanup solvent (halogenated and nonhalogenated), rags	Prepare only the quantity of ink needed for a press run. Recycle waste ink and solvent. Schedule runs to reduce color change over. Use automatic cleaning equipment. Use automatic ink leveler. Use alternative solvents. Use water-based ink. Use UV-curable ink. Install web break detectors. Use automatic web splicers. Store ink properly. Standardize ink sequence. Recycle waste ink.

Table 3. (Continued)

Waste Origin/Type	Pollution Prevention and Recycling Method
Printing/ Test production, bad printings, empty ink containers, used blankets.	Install web break detectors. Monitor press performance. Use better operating practices.
Printing/ (Continued)	Use alternative fountain solutions. Use alternative cleaning solvents. Use automatic blanket cleaners. Improve cleaning efficiency. Collect and reuse solvent. Recycle lube oils.
Finishing/ Paper waste from damaged product	Reduce paper use. Recycle waste paper.

Table 4. Checklist for the Fabricated Metal Industry

Waste Origin/Type	Pollution Prevention and Recycling Methods
Machining Wastes/ Metalworking Fluid	Use of high-quality metalworking fluid. Use demineralized water makeup. Perform regularly scheduled sump and machine cleaning. Perform regularly scheduled gasket, wiper, and seal maintenance. Filter, pasteurize, and treat metalworking fluid for reuse. Assigning fluid control responsibility to one person. Standardize oil types used on machining equipment. Improve equipment scheduling/establish dedicated lines. Reuse or recycle cutting, cooling, and lubricating oils. Substitute insoluble borates for soluble borate lubricants.
Machining Wastes/ Metal wastes, dust, and sludge	Segregate and reuse scrap metal.
Parts Cleaning/ Solvents	Install lids/silhouettes on tanks. Increase freeboard space on tanks. Install freeboard chillers on tanks. Remove sludge from solvent tanks frequently. Extend solvent life by precleaning parts by wiping, using air blowers, or predipping in cold mineral spirits dip. Reclaim/recover solvent on- or off-site. Substitute less hazardous solvent degreasers (e.g., petroleum solvents instead of chlorinated solvents) or alkali washes where possible. Distribute parts on rack to allow good cleaning and minimize solvent holup. Slow speed of parts removal from vapor zone. Rotate parts to allow condensed solvent drop-off.
Parts Cleaning/ Aqueous Cleaners	Remove sludge frequently. Use dry cleaning and stripping methods. Use oil separation and filtration to recycle solution.
Parts Cleaning/ Abrasives	Use of greaseless or water-based binders. Use an automatic liquid spray system for application of abrasive onto wheel. Ensure sufficient water use during cleaning by using water level control. Use synthetic abrasives.
Parts Cleaning/ Rinsewater	Improve rack and barrel system design. Use spray, fog, or chemical rinses. Use deionized water makeup to increase solution life.

Table 4. (Continued)

Waste Origin/Type	Pollution Prevention and Recycling Methods
Surface Treatment and Plating/ Process Solutions	Use material or process substitution e.g., trivalent chromium. Use low solvent paint for coating. Use mechanical cladding and coating. Use cleaning baths as pH adjusters. Recover metals from process solutions.

Surface Treatment and Plating/ Rinsewater	Reduction in drag-out of process chemicals:
	    Reduce speed of withdrawal
	    Lower plating bath concentrations
	    Reuse rinsewater
	    Use surfactants to improve drainage
	    Increase solution temperature to reduce viscosity
	    Position workpiece to minimize solution holdup
	System design considerations:
	    Rinsetank design
	    Multiple rinsing tanks
	    Conductivity measurement to control rinse water flow
	    Fog nozzles and sprays
	    Automatic flow controls
	    Rinse bath agitation
	    Counter current rinse.

Table 5. Checklist for the Metal Casting Industry

Waste Origin/Type	Pollution Prevention and Recycling Methods
Baghouse Dust and Scrubber Waste/ Dust contaminated with lead, zinc, and cadmium	Identify the source of contaminants, e.g., coatings on scrap, and work with suppliers to find raw materials that reduce the contaminant input. Install induction furnaces to reduce dust production. Recycle dust to original process or to another process. Recover contaminants with pyrometallurgical treatment, rotary kiln, hydrogen reduction, or other processes. Recycle to cement manufacturer.
Production of Ductile Iron/ Hazardous slag	Reduce the amount of sulfur in the feedstock. Use calcium oxide or calcium fluoride to replace calcium carbide as the desulfurization agent. Improve process control. Recycle calcium carbide slag.
Casting/ Spent casting sand	Material substitution, e.g., olivine sand is more difficult to detoxify than silica sand. Separate sand and shot blast dust. Improve metal recovery from sand. Recover sand and mix old and new sand for mold making. Recover sand by washing, air scrubbing, or thermal treatment. Reuse sand for construction if possible.

Table 6. Checklist for the Printed Circuit Board Industry

Waste Origin/Type	Pollution Prevention and Recycling Methods

PC Board Manufacture/ General	Product substitution:
	    Surface mount technology
	    Injection molded substrate and additive plating

Cleaning and Surface Preparation/ Solvents	Materials substitution:
	    Use abrasives
	    Use nonchelated cleaners
	Increase efficiency of process:
	    Extend bath life, improve rinse efficiency, countercurrent cleaning
	Recycle/reuse:
	    Recycle/reuse cleaners and rinses

Pattern Printing and Masking/ Acid fumes/organic vapors; vinyl polymers spent resist removal solution; spent acid solution; waste rinse water	Reduce hazardous nature of process:
	    Aqueous processable resist
	    Screen printing versus photolithography
	    Dry photoresist removal
	Recycle/reuse:
	    Recycle/reuse photoresist stripper

Electroplating and Electroless Plating/ Plating solutions and rinse wastes	Eliminate process:
	    Mechanical board production
	Materials substitution:
	    Noncyanide baths
	    Noncyanide stress relievers
	Extend bath life; reduce drag-in:
	    Proper rack design/maintenance, better precleaning/rinsing, use of demineralized water as makeup, proper storage methods
	Extend bath life; reduce drag-out:
	    Minimize bath chemical concentration, increase bath temperature, use wetting agents, proper positioning on rack, slow withdrawal and sample drainage, computerized/automated systems, recover drag-out, use airstreams or fog to rinse plating solution into the tank, collect drips with drain boards.
	Extend bath life; maintain bath solution quality:
	    Monitor solution activity
	    Control temperature
	    Mechanical agitation
	    Continuous filtration/carbon treatment
	    Impurity removal
	Improve rinse efficiency:
	    Closed-circuit rinses
	    Spray rinses
	    Fog nozzles

Waste Origin/Type	Pollution Prevention and Recycling Methods

Electroplating and Electroless Plating/ (Continued)	Improve rinse efficiency (continued):
	    Increased agitation
	    Countercurrent rinsing
	    Proper equipment design/operation
	    Deionized water use.
	Turn off rinsewater when not in use.
	Recovery/reuse:
	    Segregate streams
	    Recover metal values.

Etching/ Etching solutions and rinse wastes	Eliminate process:
	    Differential plating
	    Use dry plasma etching.
	Materials substitution:
	    Nonchelated etchants
	    Nonchrome etchants.
	Increased efficiency:
	    Use thinner copper cladding
	    Pattern vs. panel plating
	    Additive vs. subtractive method.
	Reuse/recycle:
	    Reuse/recycle etchants.

Table 7. Checklist for the Coating Industry

Waste Origin/Type	Pollution Prevention and Recycling Methods
Coating Overspray/ Coating material that fails to reach the object being coated	Maintain 50% overlap between spray pattern. Maintain 6- to 8-inch distance between spray gun and the workpiece. Maintain a gun speed of about 250 feet/minute. Hold gun perpendicular to the surface. Trigger gun at the beginning and end of each pass. Properly train operators. Use robots for spraying. Avoid excessive air pressure for coating atomization. Recycle overspray. Use electrostatic spray systems. Use turbine disk or bell or air-assisted airless spray guns in place of air-spray guns. Install on-site paint mixers to control material usage. Inspect parts before coating.
Stripping Wastes/ Coating removal from parts before applying a new coat	Avoid adding excess stripper. Use spent stripper as rough prestrip on next item. Use abrasive media paint stripping. Use plastic media bead-blasting paint stripping. Use cryogenic paint stripping. Use thermal paint stripping. Use wheat starch media blasting paint stripping. Use laser or flashlamp paint stripping.
Solvent Emissions/ Evaporative losses from process equipment and coated parts	Keep solvent soak tanks away from heat sources. Use high-solids coating formulations. Use powder coatings. Use water-based coating formulations. Use UV cured coating formulations.
Equipment Cleanup Wastes/ Process equipment cleaning with solvents	Use light-to-dark batch sequencing. Produce large batches of similarly coated objects instead of small batches of differently coated items. Isolate solvent-based paint spray booths from water-based paint spray booths. Reuse cleaning solution/solvent. Standardize solvent usage. Clean coating equipment after each use.
Source Reduction	Reexamine the need for coating, as well as available alternatives. Use longer lasting plastic coatings instead of paint.

The worksheets in this appendix were taken from the manual *Guides to Pollution Prevention: The Pharmaceutical Industry* (see Appendix G). These worksheets illustrate how personnel at a plant might customize the Pollution Prevention Worksheets in Appendix A to fit a specific industry or facility. For a full description of waste minimization assessment procedures, refer to the text of this manual.

**Case Study —
Example Pollution Prevention
Opportunity Detailed Assessment**

This study illustrates a pollution prevention assessment done by a small pharmaceutical company. This example is based on actual experience but uses fictitious names, processes, and data. The case study uses industry-specific worksheets and covers detailed assessment activities from forming an assessment team through screening options.

The ABC Pharmaceutical Company, Inc., is a small production facility. Its main product is a low-volume, high-value-added protein solution product. ABC also manufactures a high-volume, low-value-added saline solution product. The growing cost of waste disposal and the small margin of profit on the saline solution product led management to institute a pollution prevention program.

A pollution prevention task force was assembled. It consisted of:
- A process engineer
- A product engineer
- A process area supervisor
- An environmental compliance specialist

The process engineer was the team leader and the corporate pollution prevention champion.

The team met and established the following goals:
- Achieve a significant reduction in the generation of hazardous wastes.
- Identify data sources and deficiencies and work toward developing reliable means of measuring reductions.
- Maintain product quality.
- Maintain or improve profit margin of saline solution in light of increasing waste disposal costs.

The task force assembled as much data as possible on those operations that use toxic chemicals or generate hazardous waste. This included preparing block diagrams of several key processes. They found that, aside from purchase and shipping records and regulatory reports of releases, there were few records on hazardous materials. They were unable to prepare complete mass balances for any of the key processes but were able to identify the major waste streams. The mass balances also identified additional data that would increase understanding of the process operation without extensive new data collection.

The data gathering focused on waste sources, material-handling practices, input materials, and products. The effort started with these inputs because they were the areas most likely to yield pollution prevention opportunities and because they had the most available data. The major data sources were purchasing records, waste shipment manifests, material safety data sheets, product specifications, Superfund Amendment and Reauthorization Act (SARA) reports, and conversations with the production area workers.

The team also prepared a description of the key processes in the plant (aqueous cleaning, disinfecting, venting, general housekeeping, chemical synthesis, and research and development). They then described and prioritized the waste streams.

After collecting and reviewing the plant data, the team held a brainstorming session to generate pollution prevention options. Several pollution prevention options were identified and selected for future feasibility study and possible implementation.

Worksheet Titles

Worksheet 1. Waste Sources
Worksheet 2. Waste Minimization:
Material Handling
(2a, 2b, and 2c)
Worksheet 3. Input Materials Summary
Worksheet 4. Products Summary
Worksheet 5. Option Generation:
Material Handling
Worksheet 6. Process Description
(6a, 6b, 6c, 6d, and 6e)
Worksheet 7a. Waste Stream Summary
Worksheet 7b. Waste Description
Worksheet 8. Waste Minimization:
Reuse and Recovery
Worksheet 9. Option Generation:
Process Operation
Worksheet 10. Waste Minimization:
Good Operating Practices
Worksheet 11. Waste Minimization:
Good Operating Practices

<table>
<tr><td colspan="2">Firm <u>ABC CORP</u>
Site <u>LOS ANGELES</u>
Date <u>MARCH, 1991</u></td><td colspan="2">Pollution Prevention
Assessment Worksheets

Proj. No. ___1___</td><td colspan="2">Prepared By <u>DGL</u>
Checked By <u>PEP</u>
Sheet <u>1</u> of <u>1</u> Page <u>1</u> of <u>18</u></td></tr>
</table>

WORKSHEET 1

WASTE SOURCES

Waste Source: Material Handling	Significance at Plant		
	Low	Medium	High
Off-spec materials	X		
Obsolete raw materials	X		
Obsolete products		X	
Spills and leaks (liquids)		X	
Spills (powders)	X		
Empty container cleaning	X		
Container disposal (metal)	X		
Container disposal (paper, plastic)		X	
Pipeline/tank drainage			X
Laboratory wastes		X	
Evaporative losses	X		
Other			

Waste Source: Process Operations			
Tank cleaning		X	
Container cleaning	X		
Blender cleaning	X		
Process equipment cleaning		X	

<table>
<tr><td>Firm <u>ABC CORP</u>
Site <u>LOS ANGELES</u>
Date <u>MARCH, 1991</u></td><td>Pollution Prevention
Assessment Worksheets

Proj. No. <u>1</u></td><td>Prepared By <u>DGL</u>
Checked By <u>PEP</u>
Sheet <u>1</u> of <u>1</u> Page <u>2</u> of <u>18</u></td></tr>
</table>

<table>
<tr><td>WORKSHEET
2a</td><td>WASTE MINIMIZATION:
Material Handling</td></tr>
</table>

A. GENERAL HANDLING TECHNIQUES

Are all raw materials tested for quality before being accepted from suppliers? ☒ Yes ☐ No

Describe safeguards to prevent the use of materials that may generate off-spec product: <u>REQUIRE VENDOR</u> <u>TO PROVIDE CERTIFICATE OF ANALYSIS BEFORE USE.</u>

Is obsolete raw material returned to the supplier? ☐ Yes ☒ No

Is inventory used in first-in first-out order? ☒ Yes ☐ No

Is the inventory system computerized? ☐ Yes ☒ No

Does the current inventory control system adequately prevent waste generation? ☒ Yes ☐ No

What information does the system track? <u>PRODUCT/MATERIAL NAME, LOT NUMBER,</u> <u>CONTROL NUMBER, DATE RECEIVED, EXPIRATION DATE.</u>

Is there a formal personnel training program on raw material handling, spill prevention, proper storage techniques, and waste handling procedures? ☐ Yes ☒ No

Does the program include information on the safe handling of the types of drums, containers and packages received? ☐ Yes ☐ No

How often is training given and by whom? <u>N/A</u>

Is dust generated in the storage area during the handling of raw materials? ☐ Yes ☒ No

If yes, is there a dedicated dust recovery system in place? ☐ Yes ☐ No

Are methods employed to suppress dust or capture and recycle dust? ☐ Yes ☐ No

Explain: <u>N/A</u>

<table>
<tr><td>Firm <u>ABC CORP</u>
Site <u>LOS ANGELES</u>
Date <u>MARCH, 1991</u></td><td>Pollution Prevention
Assessment Worksheets

Proj. No. <u>1</u></td><td>Prepared By <u>DGL</u>
Checked By <u>PEP</u>
Sheet <u>1</u> of <u>1</u> Page <u>3</u> of <u>18</u></td></tr>
</table>

<table>
<tr><td>WORKSHEET
2b</td><td>WASTE MINIMIZATION:
Material Handling</td></tr>
</table>

B. BULK LIQUIDS HANDLING

What safeguards are in place to prevent spills and avoid ground contamination during the transfer and filling of storage and blending tanks?

High level shutdown/alarms	☐	Secondary containment	☑
Flow totalizers with cutoff	☐	Other	☐

Describe the system: <u>UNDERGROUND TANKS ARE LOCATED IN</u>
<u>CONCRETE PITS; ABOVE GROUND TANKS ARE SURROUNDED</u>
<u>WITH BERMS</u>

Are air emissions from solvent storage tanks controlled by means of:

Conservation vents	☐	Absorber/Condensor	☐	Adsorber	☐
Nitrogen Blanketing	☐	Other vapor loss control system	☐		

Describe the system: <u>NO SYSTEM</u>

Are all storage tanks routinely monitored for leaks? If yes, describe procedure and monitoring frequency for above-ground/vaulted tanks: <u>VISUAL INSPECTION WEEKLY</u>

Underground tanks: <u>VISUAL INSPECTION WEEKLY</u>

How are the liquids in these tanks dispensed to the users? (i.e., in small containers or hard-piped.) <u>HARD-PIPED TO MANUFACTURING AREAS.</u>

What measures are employed to prevent the spillage of liquids being dispensed? <u>EMPLOYEE TRAINING</u>

Are pipes cleaned regularly? Also discuss the way pipes are cleaned and how the resulting waste is handled: <u>N/A</u>

When a spill of liquid occurs in the plant, what cleanup methods are employed (e.g., wet or dry)? Also discuss the way in which the resulting wastes are handled: <u>WET METHOD FOR LARGE SPILLS - SQUEEGEE TO DRAIN; DRY METHOD FOR SMALL SPILLS - TOWELS TO TRASH</u>

Would different cleaning methods allow for direct reuse or recycling of the waste? (explain) <u>REUSE NOT ALLOWED PER GMP'S VOLUME TOO SMALL FOR RECYCLING</u>

<table>
<tr><td>Firm <u>ABC CORP</u>
Site <u>LOS ANGELES</u>
Date <u>MARCH, 1991</u></td><td>Pollution Prevention
Assessment Worksheets

Proj. No. <u>1</u></td><td>Prepared By <u>DGL</u>
Checked By <u>DEP</u>
Sheet <u>1</u> of <u>1</u> Page <u>4</u> of <u>18</u></td></tr>
</table>

<table>
<tr><td>WORKSHEET
2c</td><td>WASTE MINIMIZATION:
Material Handling</td></tr>
</table>

C. DRUMS, CONTAINERS, AND PACKAGES

Are drums, containers, and packages inspected for damage before being accepted? ☒ Yes ☐ No

Are employees trained in ways to safely handle the types of drums and packages received? ☒ Yes ☐ No

Are they properly trained in handling of spilled materials? ☒ Yes ☐ No

Are stored items protected from damage, contamination, or exposure to rain, snow, sun and heat? ☒ Yes ☐ No

Describe handling procedures for damaged items: <u>PLACE IN LARGE PLASTIC BAG, RETURN TO DISTRIBUTOR</u>

Does the layout of the plant result in heavy traffic through the raw material storage area? (Heavy traffic increases the potential for contaminating raw materials with dirt or dust and for causing spilled materials to become dispersed throughout the facility.) ☐ Yes ☒ No

Can traffic through the storage area be reduced? *N/A* ☐ Yes ☐ No

To reduce the generation of empty bags and packages, dust from dry material handling and liquid wastes due to cleaning empty drums, has the plant attempted to:

Purchase hazardous materials in preweighed containers to avoid the need for weighing? ☐ Yes ☒ No

Use reusable/recyclable drums with liners instead of paper bags? ☒ Yes ☐ No

Use larger containers or bulk delivery systems that can be returned to supplier for cleaning? ☐ Yes ☒ No

Dedicate systems in the loading area so as to segregate hazardous from nonhazardous wastes? ☒ Yes ☐ No

Recycle the cleaning waste into a product? ☐ Yes ☒ No

Describe the results of these attempts: <u>DRUMS WITH LINERS O.K.</u>

Are all empty bags, packages, and containers that contained hazardous materials segregated from those that contain nonhazardous wastes? Describe the method currently used to dispose of this waste: <u>NO HAZARDOUS WASTES GENERATED BY PLANT, LAB HAZARDOUS WASTES PUT IN LAB PACKS AND REMOVED FROM FACILITY</u>

<table>
<tr><td>Firm <u>ABC CORD</u>
Site <u>LOS ANGELES</u>
Date <u>MARCH, 1991</u></td><td>Pollution Prevention
Assessment Worksheets

Proj. No. <u>1</u></td><td>Prepared By <u>DGL</u>
Checked By <u>PEP</u>
Sheet <u>1</u> of <u>1</u> Page <u>5</u> of <u>18</u></td></tr>
</table>

WORKSHEET
3

INPUT MATERIALS SUMMARY

Attribute	Description		
	Stream No. 1	Stream No. 1	Stream No. ___
Material Name/ID	SODIUM CHLORIDE	WATER (WFI)	
Source/Supplier	JT BAYER	MADE IN HOUSE	
Hazardous Component	N/A	N/A	
Annual Consumption Rate	50,000 Kg	1,000,000 L	
Purchase Price, $ per ______	$5/Kg	$4/L	
Overall Annual Cost	$250,000	$4,000,000	
Material Flow Diagram available (Y/N)	N	Y	
Delivery Mode[1]	TRUCK	PIPELINE	
Shipping Container Size and Type[2]	25 LB DRUM	N/A	
Storage Mode[3]	WAREHOUSE	N/A	
Transfer Mode[4]	HAND TRUCK	PUMP	
Control Mode[5]	SIGN OUT	AVAIL. TO ALL	
Empty Container Disposal Management[6]	CRUSH & LANDFILL	N/A	
Shelf Life	2 YR	N/A	
Supplier Would			
— accept expired material? (Y/N)	Y	N/A	
— accept shipping containers? (Y/N)	Y	N/A	
— revise expiration date? (Y/N)	Y DOWN GRADE	N/A	
Acceptable Substitute(s), if any	NONE	NONE	
Alternate Supplier(s)	SEVERAL		

Notes:
1. e.g., pipeline, tank car, 100 bbl tank truck, truck, etc.
2. e.g., 55 gal drum 100 lb paper bag, tank, etc.
3. e.g., outdoor, warehouse, underground, aboveground, etc.
4. e.g., pump, forklift, pneumatic transport, conveyor, etc.
5. e.g., on-demand to all, select people only, sign out.
6. e.g., crush and landfill, clean and recycle, return to supplier, etc.

<table>
<tr><td>Firm <u>ABC CORP</u>
Site <u>LOS ANGELES</u>
Date <u>MARCH, 1991</u></td><td>Pollution Prevention
Assessment Worksheets

Proj. No. <u>1</u></td><td>Prepared By <u>DGL</u>
Checked By <u>PEP</u>
Sheet <u>1</u> of <u>1</u> Page <u>6</u> of <u>18</u></td></tr>
</table>

WORKSHEET 4

PRODUCTS SUMMARY

Attribute	Description		
	Stream No. <u>1</u>	Stream No. <u>2</u>	Stream No. ___
Name/ID	SALINE SOLUTION	PROTEIN SOLUTION	
Hazardous Component	—	—	
Annual Production Rate	1,000,000 L	3,000 L	
Annual Revenues, $ ___	$8.5 MILLION	$230 MILLION	
Shipping Mode	TRUCK	TRUCK	
Shipping Container Size and Type	CARDBOARD BOX 3'x2'x1'	CARDBOARD BOX 3'x2'x1'	
On-site Storage Mode	WAREHOUSE	COLD STORAGE	
Containers Returnable (Y/N)	N	N	
Shelf Life	1 YEAR	6 MOS	
Re-work Possible (Y/N)	Y	N	
Customer would:			
— Relax specification (Y/N)	N	N	
— Accept larger containers (Y/N)	N	N	

Appendix C

<table>
<tr><td colspan="2">Firm <u>ABC CORP</u>
Site <u>LOS ANGELES</u>
Date <u>MARCH, 1991</u></td><td>**Pollution Prevention**
Assessment Worksheets

Proj. No. <u>1</u></td><td>Prepared By <u>DGL</u>
Checked By <u>DEP</u>
Sheet <u>1</u> of <u>1</u> Page <u>7</u> of <u>8</u></td></tr>
</table>

<table>
<tr><td>WORKSHEET
5</td><td>**OPTION GENERATION:**
Material Handling</td></tr>
</table>

Meeting Format (e.g., brainstorming, nominal group technique) <u>BRAINSTORMING</u>
Meeting Coordinator <u>DGL</u>
Meeting Participants <u>MAT, DEP, DLS</u>

Suggested Waste Minimization Options	Currently Done Y/N?	Rationale/Remarks on Option
A. General Handling Techniques		
Quality Control Check	Y	
Return Obsolete Material to Supplier	N	SUPPLIER WOULD TAKE BACK AND DOWNGRADE
Minimize Inventory	Y	
Computerize Inventory	N	NOT COST EFFECTIVE
Formal Training	N	TO ESTIMATE LOST/BENEFIT
B. Bulk Liquids Handling		
High Level Shutdown/Alarm	Y	
Flow Totalizers with Cutoff	N	
Secondary Containment	Y	
Air Emission Control	N	LOOK INTO THIS
Leak Monitoring	N	COST ??
Spilled Material Reuse	N	AGAINST GMP
Cleanup Methods to Promote Recycling	N	EXAMINE SPILL TYPES
C. Drums, Containers, and Packages		
Raw Material Inspection	Y	
Proper Storage/Handling	Y	
Preweighed Containers	N	NEED TO WEIGH & VERIFY IN HOUSE
Soluble Bags	N	NO
Reusable Drums	Y	
Bulk Delivery	N	TO BE CONSIDERED
Waste Segregation	N	NO HAZARDOUS MANUFACTURED WASTE
Reformulate Cleaning Waste	N	

<table>
<tr><td>Firm <u>ABC CORP</u>
Site <u>LOS ANGELES</u>
Date <u>MARCH 1991</u></td><td>Pollution Prevention
Assessment Worksheets

Proj. No. _____1_____</td><td>Prepared By <u>DGL</u>
Checked By <u>PEP</u>
Sheet <u>1</u> of <u>1</u> Page <u>8</u> of <u>18</u></td></tr>
</table>

<table>
<tr><td>WORKSHEET
6a</td><td>**PROCESS DESCRIPTION**</td></tr>
</table>

1. GENERAL

Aqueous Cleaning

Type of Aqueous Cleaner	Cleaning Procedure (CIP, manual wash)	Hazardous or Active Ingredient
Alkaline Sufactant	MANUAL	KOH, PHOSPHATES
Alkaline Cleaner	CIP	NaOH
Acid Cleaner		
Acid Sanitizer	CIP	H₃PO₄
Other		

How are spent cleaning solutions managed:

Biodegradable; disposed of in sewer	☑ Yes	☐ No
Treated on-site; disposed of in sewer	☑ Yes	☐ No
Transported off-site	☐ Yes	☑ No
Other	☐ Yes	☑ No

If yes, explain: <u>MANUAL WASH SOLUTION TO SEWER, CIP SOLN'S NEUTRALIZED THEN TO SEWER</u>

List waste streams generated by aqueous cleaning: <u>TANK & EQUIPMENT CLEANING</u>

Solvent Cleaning N/A

Type of Solvent Used	Cleaning Procedure	Hazardous or Active Ingredient

How are spent solutions managed:

Biodegradable; disposed of in sewer	☐ Yes	☐ No
Treated on-site; disposed of in sewer	☐ Yes	☐ No
Transported off-site	☐ Yes	☐ No
Other	☐ Yes	☐ No

If yes, explain: ____

List waste streams generated by solvent cleaning: _______________________

Appendix C

<table>
<tr><td>Firm <u>ABC CORP</u>
Site <u>LOS ANGELES</u>
Date <u>MARCH, 1991</u></td><td>Pollution Prevention
Assessment Worksheets

Proj. No. <u>1</u></td><td>Prepared By <u>DGL</u>
Checked By <u>PEP</u>
Sheet <u>1</u> of <u>1</u> Page <u>9</u> of <u>18</u></td></tr>
</table>

WORKSHEET
6b

PROCESS DESCRIPTION

1. GENERAL (continued)

Disinfecting/Sterilizing

Type of Disinfectant Used	Disinfecting Procedure (Spray, Wipedown, etc.)	Hazardous or Active Ingredient
BLUE CHIP	WIPEDOWN	QUAT ANMONIUM CMPDC

How are spent disinfectants managed:

Biodegradable; disposed of in sewer   ☑ Yes ☐ No
Treated on-site; disposed of in sewer   ☐ Yes ☐ No
Transported off-site   ☐ Yes ☐ No
Other   ☐ Yes ☐ No

If yes, explain: <u>EXCESS SOLUTION DUMPED TO DRAIN</u>

Is ethylene oxide used for sterilization?   ☐ Yes ☑ No

What type of pollution control equipment is used? _______________

What is the percent (%) ethylene oxide captured? _______________

What is the percent (%) chlorofluorocarbon captured? _______________

List waste streams generated by disinfecting/sterilizing: <u>EXCESS SOLUTION FROM DAILY FILL AREA WIPEDOWN</u>

Venting

What large-volume liquid chemicals are stored on-site? <u>ETHANOL</u>

Are storage tanks with breathing vents used? <u>YES</u>

Do process vessels release vapors? <u>YES</u>

What chemicals are released through vessel vents? <u>ETHANOL</u>

What type of pollution control equipment is in place? <u>NONE</u>

What percent (%) of vent gases generated are captured? <u>Ø</u>

List waste streams generated by venting: <u>ETHANOL VAPOR</u>

| WORSHEET **6c** | **PROCESS DESCRIPTION** |

1. GENERAL (continued)

Disposables

List the disposable items used in manufacturing: _PLASTIC PIPETS, BEAKERS, ETC_

Off-Spec Materials

List the production raw materials that have been disposed of due to being out-dated or off-spec: _SODIUM CHLORIDE_

List the products you manufacture that have been destroyed and disposed of due to being out-dated or off-spec: _SALINE SOLUTION_

How are these items managed? _NaCl TO MANUFACTURER, SOMETIMES TRASH. SOLUTIONS ARE REWORKED_

2. FERMENTATION _N/A_

Fermenter Information

Description of fermenter: ____

Identification number: ____

Type of growth media used: ____

Size of sump: ____

Frequency of sump cleanout: ____

Does sump fluid go to waste treatment tank? ___________________________

How often is fermenter inspected for the following:

    Heat transfer fluid leakage: _______________________________________

    Agitator seal fluid leakage: _______________________________________

    Integrity of process connectors: ___________________________________

    Integrity of sterile barriers: _____________________________________

What is the length of the fermentation cycle? _________________________

Process Information

How is culture removed from fermenter? _________________________________

____

____

| Firm _ABC CORP_
 Site _LOS ANGELES_
 Date _MARCH, 1991_ | **Pollution Prevention**
 Assessment Worksheets

 Proj. No. ____1____ | Prepared By _DGL_
 Checked By _PEP_
 Sheet _1_ of _1_ Page _11_ of _58_ |

WORKSHEET 6d

> **PROCESS DESCRIPTION**

2. FERMENTATION (continued)

Where does it go? ______________________________

How are cells removed? ______________________________

Is used media sterilized? ________ If so, how: ______________________________

Are media, cell debris, or vent gas waste streams hazardous? ______________________________

If yes, list hazardous components: ______________________________

How are contaminated fermentation batches handled? ______________________________

What is the fermentation yield percentage? ______________________________

List the waste streams that are generated by fermentation: ______________________________

3. CHEMICAL SYNTHESIS, NATURAL PRODUCT EXTRACTION, FORMULATION

Solvent-Based Processes

Solvent	Operation	Annual Usage
ETHANOL	PROTEIN SEPARATIONS	1,600,000 GAL

How are spent solvents managed: _RECOVER AND RECLAIM BY DISTILLATION_

List waste streams generated by solvent-based processes: _SPENT ETHANOL SOLUTION CONTAINING DENATURED PROTEINS_

<table>
<tr><td>Firm <u>ABC CORP</u>
Site <u>LOS ANGELES</u>
Date <u>MARCH, 1991</u></td><td>Pollution Prevention
Assessment Worksheets

Proj. No. <u>1</u></td><td>Prepared By <u>DGL</u>
Checked By <u>PEP</u>
Sheet <u>1</u> of <u>1</u> Page <u>12</u> of <u>18</u></td></tr>
</table>

WORKSHEET
6e

PROCESS DESCRIPTION

3. CHEMICAL SYNTHESIS, NATURAL PRODUCT EXTRACTION, FORMULATION (continued)

Aqueous-Based Processes

What types of water are used in your plant?

Water for injection	☑ Yes	☐ No
Distilled water	☑ Yes	☐ No
Softened water	☐ Yes	☐ No
Municipal water	☑ Yes	☐ No
Reverse osmosis/Deionized water	☐ Yes	☐ No

What aqueous process solutions are generated or used?

Aqueous Solution	Type of Water	Operation	Annual Usage
SODIUM CHLORIDE	WFI	FORMULATION	1,000,000 L
lab REAGENTS	DISTILLED H₂O	LAB OPERATIONS	UNKNOWN

How are spent cleaning solutions managed:

Biodegradable; disposed of in sewer	☑ Yes	☐ No
Recycled on-site	☑ Yes	☐ No
Recycled off-site	☐ Yes	☐ No
Treated on-site	☐ Yes	☐ No
Treated off-site	☐ Yes	☐ No
Other	☐ Yes	☐ No

If yes, explain: <u>SALINE SOLUTIONS RECYCLED ON-SITE IF OUT OF SPEC,
LAB SOLUTIONS ARE PUT DOWN THE DRAIN</u>

List waste streams generated by aqueous-based processes: <u>SPENT LABORATORY REAGENTS</u>

4. RESEARCH AND DEVELOPMENT

List disposable items used in R&D processes: <u>MISC PLASTIC LABWARE</u>

List other R&D wastes:

Process	Type of Waste	Current Waste Management Method
FILTRATION	FILTER MEDIA	TRASH/LANDFILL
PACKAGING WASTE	PAPER, PLASTICS	TRASH/LANDFILL

Appendix C

<table>
<tr><td colspan="2">Firm <u>ABC CORP</u>
Site <u>LOS ANGELES</u>
Date <u>MARCH, 1991</u></td><td colspan="2">Pollution Prevention
Assessment Worksheets

Proj. No. <u>1</u></td><td colspan="2">Prepared By <u>DGL</u>
Checked By <u>PEP</u>
Sheet <u>1</u> of <u>1</u>   Page <u>13</u> of <u>18</u></td></tr>
</table>

WORKSHEET
7a

WASTE STREAM SUMMARY

Attribute	Relative[3] Wt. (W)	Stream No. 1 Rating (R)	R x W	Stream No. 2 Rating (R)	R x W	Stream No. 3 Rating (R)	R x W
Waste ID/Name:		CLEANING WASTE		ETOH VAPOR		PKG. WASTE	
Source/Origin		TANK AND EQUIPMENT CLEANING		STORAGE TANK VENTS		WASTE PACKAGING MATERIALS	
Annual Generation Rate (units/year)		150,000 GAL/YR		1000 GAL/YR		25,000 LB/YR	
Hazardous Component Name		—		—		—	
Annual Rate of Component(s) of Concern		—		—		—	
Annual Cost of Disposal		$1000		$1000		$10,000	
Unit Cost ($/______)		$0.0067/GAL		$1/GAL		$0.4/LB	
Method of Management[1]		SEWER		AIR EMISSION		SANITARY LANDFILL	
Priority Rating Criteria[2]							
Regulatory Compliance	8	5	40	9	72	4	32
Treatment/Disposal Cost	4	3	12	3	12	7	28
Potential Liability	7	5	35	5	35	5	35
Waste Quantity Generated	6	9	54	4	24	6	36
Waste Hazard	2	2	4	6	12	2	4
Safety Hazard	3	1	3	1	3	1	3
Minimization Potential	5	8	40	7	35	8	40
Potential to Remove Bottleneck	2	2	4	2	4	2	4
Potential By-product Recovery	0	1	0	1	0	1	0
Sum of Priority Rating Scores		Σ(RxW)	192	Σ(RxW)	197	Σ(RxW)	182
Priority Rank			2		1		3

Notes: 1. For example, sanitary landfill, hazardous waste landfill, on-site recycle, incineration, etc.

2. Rate each stream in in each category on a scale from 0 (none) to 10 (high).

3. A very important criteria for your plant would receive a weight of 10; a relatively unimportant criteria might be given a weight of 2 or 3.

<table>
<tr><td>Firm _ABC CORP_
Site _LOS ANGELES_
Date _MARCH, 1991_</td><td>**Pollution Prevention
Assessment Worksheets**

Proj. No. ___1___</td><td>Prepared By __DGL__
Checked By __PEP__
Sheet _1_ of _1_ Page _14_ of _18_</td></tr>
</table>

WORKSHEET
7b

WASTE DESCRIPTION

1. Waste Stream Name/ID: _ETHANOL VAPOR_ Stream # _2_
Process Unit/Operation _STORAGE TANK EMISSIONS_

2. Waste characteristics (attach additional sheet with composition data, as necessary)
☒ gas ☐ liquid ☐ solid ☐ mixed phase
Density, lb/cu. ft. _______________ High Heating Value, Btu/lb _______________
Viscosity/Consistency _______________
pH _______________ flash point _______________ % water _______________

3. Waste leaves process as:
☒ air emission ☐ waste water ☐ solid waste ☐ hazardous waste
☐ other _______________
____

4. Waste generation is:
☒ continuous _HIGHER IN SUMMERTIME DUE TO HIGH TEMPERATURE_
☐ discrete _______________
discharge triggered by: ☐ chemical analysis _______________
☒ other (describe) _TANK VENTING_

Type: ☐ periodic _______ length of period:
☐ sporadic (irregular occurrence)
☐ non-recurrent _______________

5. Generation Rate
Annual _1,000_ ~~lbs~~ GAL per year _______________
Maximum _______________ lbs per year _______________
Average _______________ lbs per year _______________
Frequency _______________ batches per _______________
Batch Size _______ Average _______ Range _______________

6. Waste Origins/Sources
(Fill out this worksheet to identify the origin of the waste. If the waste is a mixture of waste streams, fill out a sheet for each of the individual wastes).

Is waste mixed with other wastes? ☐ yes ☒ no
Is waste segregation possible? ☐ yes ☒ no

If yes, what can be segregated from it? _______________
____

If no, why not? _PURE VAPOR_
Input material source of this waste _ETHANOL STORAGE_
____

  Appendix C

<table>
<tr><td>Firm ABC CORP
Site LOS ANGELES
Date MARCH, 1991</td><td>Pollution Prevention
Assessment Worksheets

Proj. No. 1</td><td>Prepared By DGL
Checked By DEP
Sheet 1 of 1 Page 15 of 18</td></tr>
</table>

<table>
<tr><td>WORKSHEET
8</td><td>WASTE MINIMIZATION:
Reuse and Recovery</td></tr>
</table>

A. SEGREGATION   N/A

Segregation of wastes reduces the amount of unknown material in waste and improves prospects for reuse and recovery.

Are different solvent wastes from equipment cleanup segregated?	☐ Yes	☐ No
Are aqueous wastes from equipment cleanup segregated from solvent wastes?	☐ Yes	☐ No
Are spent alkaline solutions segregated from the rinse water streams?	☐ Yes	☐ No

If no, explain: ____

____

B. ON-SITE RECOVERY

On-site recovery of solvents by distillation is economically feasible for as little as 8 gallons of solvent waste per day.

Has on-site distillation of the spent solvent ever been attempted?	☒ Yes	☐ No
If yes, is distillation still being performed?	☒ Yes	☐ No

If no, explain: ____

____

C. CONSOLIDATION/REUSE

Are many different solvents used for cleaning?	☐ Yes	☒ No
If too many small-volume solvent waste steams are generated to justify on-site distillation, can the solvent used for equipment cleaning be standardized?	☐ Yes	☐ No
Is spent cleaning solvent reused?	☐ Yes	☐ No
Are there any attempts at making the rinse solvent part of a batch formulation (rework)?	☐ Yes	☐ No
Are any attempts made to blend various waste streams to produce marketable products?	☐ Yes	☐ No
Are spills collected and reworked?	☐ Yes	☒ No

Describe which measures have been successful: ____

____

Is your solvent waste segregated from other wastes?	☒ Yes	☐ No
Has off-site reuse of wastes through waste exchange services been considered?	☐ Yes	☒ No
Or reuse through commercial brokerage firms?	☐ Yes	☒ No

If yes, results: ____

____

____

<table>
<tr><td>Firm ABC CORP
Site LOS ANGELES
Date MARCH, 1991</td><td>Pollution Prevention
Assessment Worksheets

Proj. No. 1</td><td>Prepared By DGL
Checked By PEP
Sheet 1 of 1 Page 16 of 18</td></tr>
</table>

<table>
<tr><td>WORKSHEET
9</td><td>OPTION GENERATION:
Process Operation</td></tr>
</table>

Meeting Format (e.g., brainstorming, nominal group technique) BRAINSTORMING

Meeting Coordinator DGL

Meeting Participants MAT, PEP, DLS, MJS

Suggested Options	Currently Done Y/N?	Rationale/Remarks on Option
A. Substitution/Reformulation Options		
Solvent Substitution	N	
Product Reformulation	N	NOT FEASIBLE
Other Raw Material Substitution	N	
B. Cleaning		
Vapor Recovery	N	TO STUDY THIS FOR ETOH.
Tank Wipers	N	
Pressure Washers	Y	
Reuse Cleaning Solutions	Y	FOR CIP ONLY
Spray Nozzles on Hoses	Y	
Mop and Squeegees	Y	
Reuse Rinsewater	Y	CIP ONLY – FINAL RINSE USED AS PRERINSE
Reuse Cleaning Solvent	N	
Dedicated Equipment	Y	
Clean with Part of Batch	N	
Segregate Wastes for Reuse	N	

<table>
<tr><td>Firm <u>ABC CORP</u>
Site <u>LOS ANGELES</u>
Date <u>MARCH, 1991</u></td><td>Pollution Prevention
Assessment Worksheets

Proj. No. <u>1</u></td><td>Prepared By <u>DGL</u>
Checked By <u>PEP</u>
Sheet <u>1</u> of <u>1</u> Page <u>17</u> of <u>18</u></td></tr>
</table>

<table>
<tr><td>WORKSHEET
10</td><td>WASTE MINIMIZATION:
Good Operating Practices</td></tr>
</table>

A. PRODUCTION SCHEDULING TECHNIQUES

Is the production schedule varied to decrease waste generation? (For example, do you attempt to increase size of production runs and minimize cleaning by accumulating orders or production for inventory?)

Describe: <u>YES — SOMETIMES EQUIPMENT IS RINSED ONLY (NOT WASHED) BETWEEN RUNS THAT WILL GO TO ONE BATCH</u>

Does the production schedule include sequential formulations that do not require cleaning between batches?

If yes, indicate results: <u>SEE ABOVE</u>

Are there any other attempts at eliminating cleanup steps between subsequent batches? If yes, results:

B. AVOID OFF-SPEC PRODUCTS

Is the batch formulation attempted in the lab before large scale production? ☑ Yes ☐ No

Are laboratory QA/QC procedures performed on a regular basis? ☑ Yes ☐ No

C. OTHER OPERATING PRACTICES

Are plant material balances routinely performed? ☐ Yes ☑ No

Are they performed for each material of concern (e.g., solvent) separately? ☐ Yes ☑ No

Are records kept of individual wastes with their sources of origin and eventual disposal? ☐ Yes ☑ No
(This can aid in pinpointing large waste streams and focusing reuse efforts.)

Are the operators provided with detailed operating manuals or instruction sets? ☑ Yes ☐ No

Are all operator job functions well defined? ☑ Yes ☐ No

Are regularly scheduled training programs offered to operators? ☑ Yes ☐ No

Are there employee incentive programs related to pollution prevention? ☐ Yes ☑ No

Does the plant have an established pollution prevention program in place? ☐ Yes ☑ No

If yes, is a specific person assigned to oversee the success of the program? ☐ Yes ☐ No

Discuss goals of the program and results:

Has a pollution prevention assessment been performed at this plant in the past? If yes, discuss:
<u>NO</u>

<table>
<tr><td>

Firm __ABC CORP__
Site __LOS ANGELES__
Date __MARCH, 1991__

</td><td>

**Pollution Prevention
Assessment Worksheets**

Proj. No. ___1___

</td><td>

Prepared By __DGL__
Checked By __PEP__
Sheet _1_ of _1_ Page _18_ of _18_

</td></tr>
</table>

WORKSHEET **11**	**OPTION GENERATION:** **Good Operating Practices**

Meeting Format (e.g., brainstorming, nominal group technique) __BRAINSTORMING__

Meeting Coordinator __DGL__

Meeting Participants __MAT, PEP, HTM, JEG, DLS__

Suggested Options	Currently Done Y/N?	Rationale/Remarks on Option
A. Production Scheduling Techniques		
Increase Size of Production Run	Y	COULD GO LARGER
Sequential Formulating	Y	
Avoid Unnecessary Cleaning	Y	CLEANING USUALLY NECESSARY
Maximize Equipment Dedication	Y	
B. Avoid Off-Spec Products		
Test Batch Formulation in Lab	Y	
Regular QA/QC	Y	
C. Good Operating Practices		
Perform Material Balances	N	ENGR. TO DO THIS
Keep Records of Waste Sources & Disposition	N	PROD'N TO DO THIS
Waste/Materials Documentation	N	PROD'N TO DO THIS
Provide Operating Manuals/Instructions	Y	
Employee Training	Y	
Increased Supervision	Y	
Provide Employee Incentives	N	TO BE CONSIDERED
Increase Plant Sanitation	N	
Establish Pollution Prevention Policy	Y	ENGR/PROD'N TO DO THIS
Set Goals for Source Reduction	Y	ENGR/PROD'N TO DO THIS
Set Goals for Recycling	N	
Conduct Annual Assessments	Y	FORM WM TEAM

There are a number of organizations that can assist you in developing and maintaining a pollution prevention program. This appendix lists offices of the U.S. EPA, state agencies, and assistance programs.

U.S. ENVIRONMENTAL PROTECTION AGENCY

Pollution Prevention Information Clearinghouse

The PPIC is dedicated to reducing industrial pollutants through technology transfer, education, and public awareness. It provides technical, policy, programmatic, legislative, and financial information upon request.

The PPIC provides businesses and government agencies with information to assist them in a range of pollution prevention activities, such as:

- Establishing pollution prevention programs
- Learning about new technical options arising from U.S. and foreign R&D
- Locating and ordering documents
- Identifying upcoming events
- Discovering grant and project funding opportunities
- Identifying pertinent legislation
- Saving money by reducing waste

The PPIC disseminates this information through a number of services. These include:

- a telephone hotline
- a repository of publications, reports, and industry-specific fact sheets
- an electronic information exchange network
- indexed bibliographies and abstracts of reports, publications, and case studies
- a calendar of conferences and seminars
- a directory of waste exchanges
- information packets and workshops.

The electronic network maintained by PPIC is designated as PIES. It provides access to information databases and can be used to place orders for documents. The subsystems of PIES include:

- a message center
- a publication reference database
- a directory of experts
- case studies
- a calendar of events
- program studies
- legislation summaries
- topical mini-exchanges.

This interactive system can deliver information to the user through screen display, downloading, and FAX. It is available to off-site computers via modem 24 hours a day. For information on linking to PIES, contact:

PIES Technical Assistance
Science Applications International Corp.
8400 Westpart Drive
McLean, VA 22102
(703) 821-4800

The PPIC operates a telephone hotline for questions and requests for information. The hotline provides users who cannot access PIES electronically with access to its information and services.

For information on any of PPIC's services, write to:

U.S. EPA Pollution Prevention Office
401 M Street S.W. (PM-219)
Washington, D.C. 20460

or call:

Myles E. Morse
Office of Environmental Engineering and
  Technology Demonstration
(202) 475-7161

or:

Priscilla Flattery
Pollution Prevention Office
(202) 245-3557

Other U.S. EPA offices that can provide pollution prevention information include:

U.S. EPA Solid Waste Office
Waste Management Division
401 M Street SW
Washington, D.C. 20460
(703) 308-8402

U.S. EPA Office of Pollution Prevention and
  Toxics
401 M Street SW
Washington, D.C. 20460
(202) 260-3810

U.S. EPA Office of Air and Radiation
401 M Street SW
Washington, D.C. 20460
(202) 260-7400

U.S. EPA Office of Water
401 M Street SW
Washington, D.C. 20460
(202) 260-5700

U.S. EPA Office of Research & Development
Center for Environmental Research Information
26 Martin Luther King Drive
Cincinnati, OH 45268
(513) 569-7562

U.S. EPA Risk Reduction Engineering Laboratory
26 Martin Luther King Drive
Cincinnati, OH 45268
(513) 569-7931

U.S. EPA Office of Solid Waste and
Emergency Response
[For questions regarding RCRA and Superfund
(CERCLA), call (800) 424-9346 or
(703) 920-9810. To reach the Analytical
Hotline, call (703) 821-4789.]

U.S. EPA Regional Offices:

Region 1 (VT, NH, ME, MA, CT, RI)
John F. Kennedy Federal Building
Boston, MA 02203
(617) 565-3420

Region 2 (NY, NJ, PR, VI)
26 Federal Plaza
New York, NY 10278
(212) 264-2525

Region 3 (PA, DE, MD, WV, VA, DC)
841 Chestnut Street
Philadelphia, PA 19107
(215) 597-9800

Region 4 (KY, TN, NC, SC, GA, FL, AL, MS)
345 Courtland Street, NE
Atlanta, GA 30365
(404) 347-4727

Region 5 (WI, MN, MI, IL, IN, OH)
230 South Dearborn Street
Chicago, IL 60604
(312) 353-2000

Region 6 (NM, OK, AR, LA, TX)
1445 Ross Avenue, Suite 1200
Dallas, TX 75202
(214) 655-6444

Region 7 (NE, KS, MO, IA)
726 Minnesota Ave
Kansas City, KS 66101
  (913) 551-7050

Region 8 (MT, ND, SD, WY, UT, CO)
999 18th Street
Denver, CO 80202-2405
(303) 293-1603

Region 9 (CA, NV, AZ, HI, GU)
75 Hawthorne Street
San Francisco, CA 94105
(415) 744-1305

Region 10 (AK, WA, OR, ID)
1200 Sixth Avenue
Seattle, WA 98101
(206) 553-4973

STATE LEVEL

The following lists agencies at the state or territory level as well as universities and other organizations that can provide assistance in the areas of pollution prevention and treatment:

Alabama

Department of Environmental Management
1751 Congressman W.L. Dickenson Drive
Montgomery, AL 36130
(205) 271-7939

Environmental Institute for Waste Management
  Studies
University of Alabama
Box 870203
Tuscaloosa, AL 35487-0203
(205) 348-8403

Hazardous Material Management and Resource
  Recovery Program (HAMMAR)
University of Alabama
Tuscaloosa, AL 35487-0203
(205) 348-8401
FAX 348-9659

Retired Engineers Waste Reduction Program
P.O. Box 1010
Muscle Shoals, AL 35660
(205) 386-2807

Alaska

Alaska Health Project
Waste Reduction Assistance Program
1818 West Northern Lights, Suite 103
Anchorage, AK 99517
(907) 276-2864

Alaska Department of Environmental
  Conservation
Pollution Prevention Program
P.O. Box O
Juneau, AK 99811-1800
(907) 465-2671

Arizona

Arizona Department of Economic Planning and
  Development
1645 West Jefferson St.
Phoenix, AZ 85007
(602) 255-5705

Arizona Department of Environmental Quality
Office of Waste and Water Quality Management
2005 N. Central Ave, Room 304
Phoenix, AZ 85004
(602) 257-2380

Arkansas

Arkansas Industrial Development Commission
One State Capitol Mall
Little Rock, AR 72201
(501) 682-1121

Arkansas Department of Pollution Control
  and Ecology
Hazardous Waste Division — P.O. Box 8913
Little Rock, AR 72219-8913
(501) 570-2861

California

Bay Area Hazardous Waste Reduction Committee
  (BAHWRC)
City of Berkeley Environmental Health
2180 Milvia, Room 309
Berkeley, CA 94708
(415) 644-6510

Cal-EPA
Department of Toxic Substances Control
Alternative Technology Division
P.O. Box 806
Sacramento, CA 95812-0806
(916) 324-1807

California Conference of Directors of
  Environmental Health — Subcommittee for
  the Development of Hazardous Waste Programs
Ventura County Environmental Health
800 S. Victoria
Ventura, CA 93009
(805) 654-5039

California Environmental Business Resources
  Assistance Center
100 South Anaheim Boulevard
Suite 125
Anaheim, CA 92805
(714) 563-0135
(800) 352-5225

Central Valley Hazardous Waste Minimization
  Committee
Environmental Management Division
8475 Jackson Road, Suite 230
Sacramento, CA 95826
(916) 386-6160

Local Government Commission
909 12th Street
#205
Sacramento, CA 95814
(916) 448-1198

Pollution Prevention Program
San Diego County Department of Health Services
P.O. Box 85261
San Diego, CA 92186-5261
(619) 338-2205, -2215

Colorado

Pollution Prevention Waste Reduction Program
Colorado Department of Health
4210 E. 11th Ave.
Denver, CO 80220
(303) 320-8333

Connecticut

Bureau of Waste Management
Connecticut Department of Environmental
 Protection
18-20 Trinity Street
Hartford, CT 06106
(203) 566-8476

Connecticut Technical Assistance Program
900 Asylum Avenue, Suite 360
Hartford, CT 06105
(203) 241-0777

Delaware

Pollution Prevention Program in Dept. of Natural
 Resources & Environmental Control
89 Kings Highway
P.O. Box 1401
Dover, DE 19903
(302) 739-3822

District of Columbia

U.S. Department of Energy
Conservation and Renewable Energy
Office of Industrial Technologies
Office of Waste Reduction,
Waste Material Management Division
Bruce Cranford CE-222
Washington D.C. 20585
(202) 586-9496

Office of Recycling
D.C. Department of Public Works
2000 14th Street, NW, 8th Floor
Washington, D.C. 20009
(202) 939-7116

Florida

Hazardous Waste Reduction Management
Waste Reduction Assistance Program
Florida Dept. of Environmental Regulation
2600 Blair Stone Road
Tallahassee, FL 32399-2400
(904) 488-0300

Environmental Quality Corporation
259 Timberlane Road
Tallahassee, FL 32312-1542
(904) 386-7740

Waste Reduction Assistance Program
Florida Dept. of Environmental Regulation
2600 Blair Stone Road
Tallahassee, FL 32399-2400
(904) 488-0300

Georgia

Hazardous Waste Technical Assistance
 Program
Georgia Institute of Technology
GTRI/ESTL
151 6th Street
O'Keefe Building, Room 143
Atlanta, GA 30332
(404) 894-3806

Environmental Protection Division
Georgia Department of Natural Resources
205 Butler Street S.E. Room 1154
Atlanta, GA 30334
(404) 656-2833

Guam

Solid and Hazardous Waste Management Program
Guam EPA
IT&E Harmon Plaza Complex, Unit D-107
130 Rojas Street
Harmon, GU 96911
(671) 646-8863-5

Hawaii

Department of Planning and Economic Development
Financial Management and Assistance Branch
P.O. Box 2359
Honolulu, HI 96813
(808) 548-4617

Hawaii Department of Health
Solid and Hazardous Waste Branch
Waste Minimization
5 Waterfront Plaza, Suite 250
500 Ala Moana Blvd
Honolulu, HI 96813
(808) 586-4226

Idaho

Division of Environmental Quality
Department of Health and Welfare
1410 North Hilton Street
Boise, ID 83720-9000

Illinois

Hazardous Waste Research and Information Center
Illinois Department of Energy & Natural
  Resources
One E. Hazelwood Drive
Champaign, IL 61820
(217) 333-8940

Industrial Waste Elimination Research Center
Pritzker Department of Environmental Engineering
Illinois Institute of Technology
3201 South Dearborn
Room 103 Alumni Memorial Hall
Chicago, IL 60616
(312) 567-3535

Illinois Environmental Protection Agency
Office of Pollution Prevention
2200 Churchill Road
P.O. Box 19276
Springfield, IL 62794-9276
(217) 782-8700

Indiana

Environmental Management & Education Program
School of Civil Engineering
Purdue University
2129 Civil Engineering Building
West Lafayette, IN 47907-1284
(317) 494-5036

Indiana Department of Environmental Management
Office of Technical Assistance
P.O. Box 6015
105 South Meridian Street
Indianapolis, IN 46206-6015
(317) 232-8172

Iowa

Iowa Department of Natural Resources
Wallace State Office Building
900 East Grand Avenue
Des Moines, IA 50319-0034
(515) 281-5145

Iowa Waste Reduction Center
75 BRC
University of Northern Iowa
Cedar Falls, IA 50614-0185
(800) 422-3109
(319) 273-2079

Iowa Waste Reduction Center
University of Norther Iowa
75 Biology Research Complex
Cedar Falls, IA 50614
(319) 273-2079

Kansas

Division of Environment
Department of Health and Environment
Forbes Field, Building 740
Topeka, KS 66620
(913) 296-1535

Engineering Extension Program
Ward Hall 133
Kansas State University
Manhattan, KS 66506
(916) 532-6026

Kentucky

Waste Minimization Assessment Center
Department of Chemical Engineering
University of Louisville
Louisville, KY 40292
(502) 588-6357

Kentucky Partners
Room 312 Ernst Hall
University of Louisville
Louisville, KY 40292
(502) 588-7260

Louisiana

Department of Environmental Quality
Office of Solid and Hazardous Waste
P.O. Box 82178
Baton Rouge, LA 70884-2178
(504) 765-0355

Alternate Technologies Research and Development
Office of the Secretary
Louisiana Department of Environmental Quality
P.O. Box 44066
Baton Rouge, LA 70804
(504) 342-1254

Maine

Office of Pollution Prevention
Department of Environmental Protection
State House Station 17
Augusta, ME 04333
(207) 289-2811

Office of Waste Reduction and Recycling
Maine Waste Management Agency
State House Station 154
Augusta, ME 04333
(207) 289-5300

Maryland

Hazardous and Solid Waste Management
  Administration
Maryland Department of the Environment
2500 Broening Highway — Building 40
Baltimore, MD 21224
(301) 631-3315

Maryland Environment Service
2020 Industrial Drive
Annapolis, MD 21401
(301) 454-1941

Technical Extension Service
Engineering Research Center
University of Maryland
College Park, MD 20742
(301) 454-1941

Massachusetts

Executive Office of Environmental Affairs/
  Office of Technical Assistance
100 Cambridge Street, Room 1904
Boston, MA 02202
(617) 727-3260

Source Reduction Program
Massachusetts Department of Environmental
  Protection
1 Winter Street, 7th Floor
Boston, MA 02108
(617) 292-5870

Massachusetts Department of Environmental
  Protection
75 Grove Street
Worchester, MA 01606
(508) 792-7650

Michigan

Resource Recovery Section
Department of Natural Resources
P.O. Box 30241
Lansing, MI 48909
(517) 373-0540

Office of Waste Reduction Services
Michigan Departments of Commerce and Natural
  Resources
P.O. Box 30004
Lansing, MI 48909
(517) 335-1178

Minnesota

Minnesota Pollution Control Agency
Solid and Hazardous Waste Division
520 Lafayette Road
St. Paul, MN 55155-3898
(612) 296-6300

Minnesota Technical Assistance Program
1313 5th Street S.E., Suite 207
Minneapolis, MN 55414
(612) 627-4646
(800) 247-0015 (in Minnesota)

Minnesota Office of Waste Management
1350 Energy Lane
St. Paul, MN 55108
(612) 649-5741

Waste Reduction Institute for Training Application
Research, Inc. (WRITAR)
1313 5th Street, S.E.
Minneapolis, MN 55414
(612) 379-5995

Mississippi

Waste Reduction & Minimization Program
Bureau of Pollution Control
Department of Environmental Quality
P.O. Box 10385
Jackson, MS 39289-0385
(601) 961-5171

Mississippi Technical Assistance Program
  (MISSTAP) and Mississippi Solid Waste
  Reduction Assistance Program (MSWRAP)
P.O. Drawer CN
Mississippi State, MS 39762
(601) 325-8454

Missouri

Missouri Environmental Improvement and Energy
  Resources Authority
P.O. Box 744
325 Jefferson St.
Jefferson City, MO 65102
(314) 751-4919

Waste Management Program
Missouri Department of Natural Resources
P.O. Box 176
Jefferson City, MO 65102
(314) 751-3176

Montana

Department of Health and Environmental Sciences
Room A-206
Cogswell Building
Helena, MT 59620
(406) 444-3454

Solid and Hazardous Waste Bureau
Department of Health and Environmental Sciences
Cogswell Building
Room B-201
Helena, MT 59620
(406) 444-2821

Nebraska

Hazardous Waste Section
Nebraska Department of Environmental
  Control
P.O. Box 98922
Lincoln, NE 68509-8922
(402) 471-2186

Nevada

Nevada Small Business Development Center —
  Technical Assistance Program
Business Environmental Program
College of Business Administration, MS032
University of Nevada — Reno
Reno, NV 89557-0100
(702) 784-1717
(800) 882-3233 (Nevada only)

State Energy Conservation Program
Office of Community Services
Nevada Energy Program
Capital Complex
400 W. King
Carson City, NV 89710
(702) 687-4990

New Hampshire

New Hampshire Department of
  Environmental Services
Waste Management Division —
  Planning Bureau
6 Hazen Drive
Concord NH 03301-6509
(603) 271-2901
(603) 271-2902

New Jersey

New Jersey Hazardous Waste Facilities Siting
  Commission
Room 614
28 West State Street
Trenton, NJ 08608
(609) 292-1459
(609) 292-1026

Hazardous Waste Advisement Program
New Jersey Department of Environmental
  Protection & Energy
401 East State Street
Trenton, NJ 08625
(609) 777-0518

New Jersey Institute of Technology
Hazardous Substance Management Research
  Center
Advanced Technology Center Building
323 Martin Luther King Jr. Boulevard
University Heights
Newark, NJ 07102
(201) 596-5864

New Mexico

Economic Development Department
Bataan Memorial Building
State Capitol Complex
Santa Fe, NM 87503
(505) 827-0380

Hazardous and Radiation Waste Bureau
Environmental Improvement Division
1190 St. Francis Drive
Santa Fe, NM 87503
(505) 827-2926

New York

New York Environmental Facilities Corporation
50 Wolf Road
Albany, NY 12205
(518) 457-4222

Environmental Compliance Services
Erie County Office Building
95 Franklin Street
Buffalo, NY 14202
(716) 846-6716

North Carolina

Department of Environmental, Health, and Natural
  Resources
Pollution Prevention Pays Program
Office of Waste Reduction
3825 Barrett Drive, 3rd Floor
Raleigh, NC 27609-7221
(919) 733-7015
(919) 571-4100

Waste Reduction Resource Center
3825 Barrett Drive, Suite 300
P.O. Box 27687
Raleigh, NC 27611-7687
(919) 571-4100
(800) 476-8686

North Dakota

Environmental Health Section
State Department of Health
1200 Missouri Ave.
Bismarck, ND 58502
(701) 258-2070

Division of Waste Management
Department of Health
1200 Missouri Ave., Room 302
Bismarck, ND 58502-5520
(701) 224-2366

Ohio

Division of Solid and Infectious Waste
Attn: Pollution Prevention Section
Ohio Environmental Protection Agency
P.O. Box 1049
1800 Watermark Drive
Columbus, OH 43266-0149
(614) 644-2917

Ohio Technology Transfer Organization
  (OTTO)
Ohio Department of Development
77 South High Street, 26th Floor
Columbus, OH 43225-0330
(614) 644-4286

Ohio Department of Natural Resources
Fountain Square
Columbus, OH 43224-1387
(614) 265-6333

Ohio Environmental Protection Agency
Division of Solid and Hazardous Waste
  Management
Pollution Prevention Section
P.O. Box 1049
Columbus, OH 43266-0149
(614) 644-2917

Oklahoma

Oklahoma State Department of Health
Hazardous Waste Management Service
1000 N.E. 10th St.
Oklahoma City, OK 73117
(405) 271-5338

Hazardous Waste Management Service
Oklahoma State Department of Health
1000 Northeast 10th Street
Oklahoma City, OK 73152
(405) 271-7047

Oregon

Oregon Hazardous Waste Reduction Assistance
  Program
Department of Environmental Quality
811 Southwest Sixth Avenue
Portland, OR 97204-1390
(503) 229-5913 (6570)
800) 452-4011 (in Oregon)

Pennsylvania

Pennsylvania Technical Assistance Program
248 Calder Way, Suite 306
University Park, PA 16801
(814) 865-0427

Center of Hazardous Material Research
Subsidiary of the University of Pittsburgh Trust
320 William Pitt Way
Pittsburgh, PA 15238
(412) 826-5320
(800) 334-2467

Division of Waste Minimization and Planning
Department of Environmental Resources
P.O. Box 2064
Harrisburg, PA 17120
(717) 787-7382

Technical Specialist
PENNTAP
112 S. Burrowes Street
University Park, PA 16801
(814) 865-1914

NETAC
University of Pittsburgh Applied Research Center
615 William Pitt Way
Pittsburgh, PA 15238
(412) 826-5511

Puerto Rico

Government of Puerto Rico
Economic Development Administration
Box 362350
San Juan, PR 00936
(809) 758-4747

Rhode Island

Office of Environmental Coordination
Rhode Island Department of Environmental
  Management
83 Park Street
Providence, RI 02903
(401) 277-3434
(800) 253-2674 (in Rhode Island)

South Carolina

Center for Waste Minimization/Hazardous Waste
Department of Health and Environmental Control
2600 Bull Street
Columbia, SC 29201
(803) 734-5200

Hazardous Waste Management Research Fund
Institute of Public Affairs
4th Floor, Ganbrell Hall
University of South Carolina
Columbia, SC 29208
(803) 777-8157

Clemson University
Continuing Engineering Education Program
P.O. Drawer 1607
Clemson, SC 29633
(803) 656-4450

Sumter Technical College
South Carolina Environmental Training Center
506 N. Guignard Dr.
Sumter, SC 29150

South Dakota

Dept. of Environmental and Natural Resources
523 East Capitol
Pierre, SD 57501-3181
(605) 773-3151

Division of Environmental Regulations
Department of Water and Natural Resources
Joe Foss Building, Room 416
523 E. Capital Ave.
Pierre, SD 57501
(605) 773-3153

Tennessee

Tennessee Valley Authority
Mail Code Old City Hall Building 2f71b
Knoxville, TN 37901
(615) 632-3160

Tennessee Valley Authority
Mail Code HV2S27OC
Chattanooga, TN 37402
(615) 751-3731

Tennessee Valley Authority
1195 Antioch Pike
Nashville, TN 37219
(615) 360-1680

Waste Reduction Assistance Program
Center for Industrial Services
University of Tennessee
226 Capitol Blvd. Building
Suite 401
Nashville, TN 37219
(615) 242-2456

Texas

RENEW
Texas Water Commission
P.O. Box 13087 Capitol Station
Austin, TX 78711-7761
(512) 463-7761

Texas Technical University
P.O. Box 4679
Lubbock, TX 79409-3121
(806) 742-1413

Utah

Department of Chemical Engineering
3290 MEB
University of Utah
Salt Lake City, UT 84112
(801) 581-5763

Department of Environmental Quality
288 North 1460 West
Salt Lake City, UT 84114-4810
(801) 538-6121

Planning and Program Development
Bureau of Solid and Hazardous Waste
  Management
Utah Department of Health
P.O. Box 16690
288 North 1460 West Street
Salt Lake City, UT 84116-0690
(801) 538-6170

Utah State University
UMC 14
Logan, UT 84322
(801) 750-3227

Vermont

Vermont Department of Environmental
  Conservation
Pollution Prevention Division
103 South Main Street
Waterbury, VT 05671-0404
(802) 244-8702

Virginia

Air Pollution Control Board
P.O. Box 10089
Richmond, VA 23240
(804) 786-6035

Washington

Hazardous Waste Section
Mail Stop PV-11
P.O. Box 47600
Washington Department of Ecology
Olympia, WA 98504-7600
(206) 459-6000

West Virginia

Generator Assistance Program
Waste Management Section
West Virginia Division of Natural Resources
1356 Hansford Street
Charleston, WV 25301
(304) 348-5989

Wisconsin

Bureau of Solid Waste Management
Wisconsin Department of Natural Resources
P.O. Box 7921
101 South Webster Street
Madison, WI 53707
(608) 267-3763

Wyoming

Wyoming Department of Environmental Quality
Solid Waste Management Program
Herschler Building, 4th Floor, West Wing
122 West 25th Street
Cheyenne, WY 82002
(307) 777-7752

The Weighted Sum Method is a quantitative method for screening and ranking pollution prevention options. This method provides a means of quantifying the important criteria that affect waste management in a particular facility. This method involves three steps.

1. Determine what the important criteria are in terms of the program goals and constraints and the overall corporate goals and constraints. Example criteria are:
   * Reduction in waste quantity
   * Reduction in waste hazard (e.g., toxicity, flammability, reactivity)
   * Reduction in waste treatment/disposal costs
   * Reduction in raw material costs
   * Reduction in liability and insurance costs
   * Previous successful use within the company
   * Previous successful use in industry
   * Not detrimental to product quality
   * Low capital cost
   * Low operating and maintenance costs
   * Short implementation period with minimal disruption of plant operations

   The weights (on a scale of 0 to 10, for example) are determined for each of the criteria in relation to their importance. For example, if reduction in waste treatment and disposal costs are very important, while previous successful use within the company is of minor importance, then the reduction in waste costs is given a weight of 10 and the previous use within the company is given a weight of either 1 or 2. Criteria that are not important are not included or are given a weight of 0.

2. Each option is then rated on each criterion. Again a scale of 0 to 10 can be used (0 for low and 10 for high).

3. Finally, the rating of each option for a particular criterion is multiplied by the weight of the criterion. An option's overall rating is the sum of the products of rating times the weight of the criterion.

The options with the best overall ratings are then selected for the technical and economic feasibility analyses. Table E-1 presents an example using the Weighted Sum Method for screening and ranking options.

Table E-1. Sample Calculation Using the Weighted Sum Method

ABC Corporation has determined that reduction in waste treatment costs is the most important criterion, with a weight factor of 10. Other significant criteria include reduction in safety hazard (weight of 8), reduction in liability (weight of 7), and ease of implementation (weight of 5). Options X, Y, and Z are then each assigned effectiveness factors. For example, option X is expected to reduce waste by nearly 80%, and is given a rating of 8. It is given a rating of 6 for reducing safety hazards, 4 for reducing liability, and because it is somewhat difficult to implement, 2 for ease of implementation. The table below shows how the options are rated overall, with effectiveness factors estimated for options Y and Z.

Rating Criteria	Weight	Ratings for each option		
		X	Y	Z
Reduce treatment costs	10	8	6	3
Reduce safety hazards	8	6	3	8
Reduce liability	7	4	4	5
Ease of implementation	5	2	2	8
Sum of weight times ratings		166	122	169

From this screening, option Z rates the highest with a score of 169. Option X's score is 166 and option Y's score is 122. In this case, both option Z and option X should be selected for further evaluation because their scores are high and close to each other.

The following example presents a profitability analysis for a relatively large hypothetical pollution prevention project. This project represents the installation of a package unit that improves plant production while reducing raw material consumption and disposal costs. The analysis was done on a personal computer using a standard spreadsheet program. The salient data used in this evaluation are summarized below.

Capital Costs

- The delivered price of the equipment is quoted by the vendor at $170,000. This includes taxes and insurance.
- Materials costs (piping, wiring, and concrete) are estimated at $35,000.
- Installation labor is estimated at $25,000.
- Internal engineering staff costs are estimated at $7,000. Outside consultant and contractor costs are estimated at $15,000.
- Miscellaneous environmental permitting costs are estimated at $15,000.
- Working capital (including chemical inventotories, materials, and supplies) is estimated at $5,000.
- Startup costs are estimated by the vendor at $3,000.
- A contingency fund of $20,000 for unforeseen costs and/or overruns is included.
- Planning, design, and installation are expected to take 1 year.

Financing

- The project will be financed 60% by retained earnings and 40% by a bank loan.
- The bank loan will be repaid over 5 years of equal installments of principal plus interest at an annual percentage rate of 13%. Interest accrued during installation will be added into the total capital costs.

- All capital costs, except working capital and interest accrued during construction, will be depreciated over 7 years using the double-declining balance method, switching to the straight-line method when the charges by this method become greater.
- The marginal income tax rate is 34%.
- Escalation of all costs is assumed to be 5% per year for the life of the project.
- The firm's cost of capital is 15%.

Operating Costs and Revenues

- The pollution prevention project is estimated to decrease raw materials consumption by 300 units per year at a cost of $50 per unit. The project will not result in increased production. However, it will produce a marketable by-product to be recovered at a rate of 200 units per year and a price of $25 per unit.
- The project will reduce the quantity of hazardous waste disposed by 200 tons per year. The following items make the total unit disposal costs:

	Costs per ton of waste
Offsite disposal fees	$500
State generator taxes	10
Transportation costs	25
Other costs	25
TOTAL DISPOSAL COSTS	$560

- Incremental operating labor costs are estimated on the basis that the project is expected to require 1 hour of operator's time per 8-hour shift. There are 3 shifts per day and the plant operates 350 days per year. The wage rate for operators is $12.50 per hour.
- Operating supplies expenses are estimated at 30% of operating labor costs.

- Maintenance labor costs are estimated at 2% of the sum of the capital costs for equipment, materials, and installation. Maintenance supplies costs are estimated at 1% of these costs.
- Incremental supervision costs are estimated at 30% of the combined costs of operating and maintenance labor.
- The following overhead costs are estimated as a percentage of the sum of operating and maintenance labor and supervision costs.

Labor burden and benefit	28%
Plant overhead	25%
Headquarter overhead	20%

- Escalation of all costs is assumed to be 5% per year for the life of the project.
- The project life is expected to be 8 years.
- The salvage value of the project is expected to be zero after 8 years.

Results

The four-page printout in Figures F-1 through F-4 presents the pollution prevention project profitability spreadsheet program. Figure F-1 represents the input section of the program. Each of the numbers in the first three columns represents an input variable in the program. The righthand side of Figure F-1 is a summary of the capital requirement. This includes a calculation of the interest accrued during construction and the financing structure of the project.

Figure F-2 is a table of the revenues and operating cost items for each of the 8 years of the project's operating life. These costs are escalated by 5% each year for the life of the project.

Figure F-3 presents the annual cash flows for the project. The calculation of depreciation charges and the payment of interest and repayment of loan principal are also shown here. The calculation of the internal rate of return (IRR) and the net present value (NPV) are based on the annual cash flows. Because the project is leveraged (financed partly by a bank loan), the equity portion of the investment is used as the initial cash flow. The NPV and the IRR are calculated on this basis. The IRR calculated this way is referred to as the "return on equity."

The program is structured to present the NPV and IRR after each year of the project's operat-

ing life. In the example, after 6 years, the IRR is 19.92% and the NPV is $27,227.

Figure F-4 is a cash flow table based entirely on equity financing. Therefore, there are no interest payments or debt principal repayments. The NPV and the IRR in this case are based on the entire capital investment in the project. The IRR calculated this way is referred to as the "return on investment."

The results of the profitability analysis for this project are summarized below:

Method of Financing	IRR	NPV
60% equity/40% debt	26.47%	$84,844
100% equity	23.09%	$81,625

The IRR values are greater than the 15% cost of capital, and the NPVs are positive. Therefore, the project is attractive and should be implemented.

Pollution Prevention		started 5/22/87						CAPITAL REQUIREMENT	
Profitability Program		last changed 8/1/87							
		INPUT						**CAPITAL REQUIREMENT**	
Capital Cost Factors		**Operating Cost/Revenue Factors**							
								Construction Year	1
Capital Cost		Increased Production			Operating Labor				
Equipment	$170,000	Increased Rate, units/year	0		Operator hours/shift	1		Capital Expenditures	
Materials	$35,000	Price, $/unit	$100		Shifts/day	3		Equipment	$170,000
Installation	$25,000				Operating days/year	350		Materials	$35,000
Plant Engineering	$7,000	Marketable By-products			Wage rate, $/man-hour	$13.50		Installation	$25,000
Contractor/Engineering	$15,000	Rate, units/year	200					Plant Engineering	$7,000
Permitting Costs	$15,000	Price, $/unit	$40		Operating Supplies	30%		Contractor/Engineering	$15,000
Contingency	$20,000				(% of Operating Labor)			Permitting Costs	$15,000
Working Capital	$5,000	Decreased Raw Materials						Contingency	$20,000
Start-up Costs	$3,000	Decreased Rate, units/year	300		Maintenance Costs			Start-up Costs	$3,000
		Price, $/unit	$50		(% of Capital Costs)			Depreciable Capital	$290,000
% Equity	60%				Labor	2.00%		Working Capital	$5,000
% Debt	40%	Decreased Waste Disposal			Materials	1.00%		Subtotal	$295,000
Interest Rate on Debt, %	13.00%	Reduced Waste, tons/year	200					Interest on Debt	$14,230
Debt Repayment, years	5	Offsite Fees, $/ton	$500		Other Labor Costs			Total Capital Requiremen	$309,230
		State Taxes, $/ton	$10		(% of O&M Labor)				
Depreciation period	7	Transportation, $/ton	$25		Supervision	30.0%		Equity Investment	$185,538
Income Tax Rate, %	34.00%	Other Disposal Costs, $/ton	$25		(% of O&M Labor + Supervision)			Debt Principal	$109,462
		Total Disposal Costs, $/ton	$560		Plant Overhead	25.0%		Interest on Debt	$14,230
Escalation Rates, %	5.0%				Home Office Overhead	20.0%		Total Financing	$309,230
					Labor Burden	28.0%			
Cost of Capital (for NPV)	15.00%								

Figure F-1. Input Information and Capital Investment

REVENUE AND COST FACTORS									
Operating Year Number		1	2	3	4	5	6	7	8
Escalation Factor	1.000	1.050	1.103	1.158	1.216	1.276	1.340	1.407	1.477
INCREASED REVENUES									
Increased Production		$0	$0	$0	$0	$0	$0	$0	$0
Marketable By-products		$8,400	$8,820	$9,261	$9,724	$10,210	$10,721	$11,257	$11,820
Annual Revenue		$8,400	$8,820	$9,261	$9,724	$10,210	$10,721	$11,257	$11,820
OPERATING COST/SAVINGS									
Raw Materials		$15,750	$16,537	$17,364	$18,233	$19,144	$20,101	$21,107	$22,162
Disposal Costs		$117,600	$123,480	$129,654	$136,137	$142,944	$150,091	$157,595	$165,475
Maintenance Labor		($4,830)	($5,071)	($5,325)	($5,591)	($5,871)	($6,164)	($6,473)	($6,796)
Maintenance Supplies		($2,415)	($2,536)	($2,663)	($2,796)	($2,935)	($3,082)	($3,236)	($3,398)
Operating Labor		($14,884)	($15,628)	($16,409)	($17,230)	($18,091)	($18,996)	($19,946)	($20,943)
Operating Supplies		($4,465)	($4,688)	($4,923)	($5,169)	($5,427)	($5,699)	($5,984)	($6,283)
Supervision		($5,914)	($6,210)	($6,520)	($6,846)	($7,189)	($7,548)	($7,925)	($8,322)
Labor Burden		($7,176)	($7,535)	($7,911)	($8,307)	($8,722)	($9,158)	($9,616)	($10,097)
Plant Overhead		($6,407)	($6,727)	($7,064)	($7,417)	($7,788)	($8,177)	($8,586)	($9,015)
Home Office Overhead		($5,126)	($5,382)	($5,651)	($5,933)	($6,230)	($6,542)	($6,869)	($7,212)
Total Operating Costs		$82,134	$86,240	$90,552	$95,080	$99,834	$104,826	$110,067	$115,570

Figure F-2. Revenues and Operating Costs

RETURN ON EQUITY/RETURN ON ASSETS									
Construction Year	1								
Operating Year		1	2	3	4	5	6	7	8
Book Value	$290,000	$207,143	$147,959	$105,685	$64,257	$22,828	$0	$0	$0
Depreciation (by straight-line)		$41,429	$41,429	$41,429	$41,429	$41,429	$41,429	$0	$0
Depreciation (by doubleDB)		$82,857	$59,184	$42,274	$30,196	$18,359	$6,522	$0	$0
Depreciation		$82,857	$59,184	$42,274	$41,429	$41,429	$22,828	$0	$0
Debt Balance	$123,692	$123,692	$98,954	$74,215	$49,477	$24,738	$0	$0	$0
Interest Payment		$16,080	$12,864	$9,648	$6,432	$3,216	$0	$0	$0
Principal Repayment		$24,738	$24,738	$24,738	$24,738	$24,738	$0	$0	$0
CASH FLOWS									
Construction Year	1								
Operating Year		1	2	3	4	5	6	7	8
Revenues		$8,400	$8,820	$9,261	$9,724	$10,210	$10,721	$11,257	$11,820
+ Operating Savings		$82,134	$86,240	$90,552	$95,080	$99,834	$104,826	$110,067	$115,570
Net Revenues		$90,534	$95,060	$99,813	$104,804	$110,044	$115,546	$121,324	$127,390
- Depreciation		$82,857	$59,184	$42,274	$41,429	$41,429	$22,828	$0	$0
- Interest on Debt		$16,080	$12,864	$9,648	$6,432	$3,216	$0	$0	$0
Taxable Income		($8,403)	$23,013	$47,891	$56,943	$65,400	$92,718	$121,324	$127,390
- Income Tax		($2,857)	$7,824	$16,283	$19,361	$22,236	$31,524	$41,250	$43,313
Profit after Tax		($5,546)	$15,188	$31,608	$37,583	$43,164	$61,194	$80,074	$84,077
+ Depreciation		$82,857	$59,184	$42,274	$41,429	$41,429	$22,828	$0	$0
- Debt Repayment		$24,738	$24,738	$24,738	$24,738	$24,738	$0	$0	$0
After-Tax Cash Flow		$52,572	$49,634	$49,144	$54,273	$59,854	$84,022	$80,074	$84,077
Cash Flow for ROE	($185,538)	$52,572	$49,634	$49,144	$54,273	$59,854	$84,022	$80,074	$84,077
Net Present Value	($185,538)	($139,823)	($102,293)	($69,980)	($38,949)	($9,191)	$27,134	$57,237	$84,722
Return on Equity		-71.66%	-32.21%	-9.63%	4.22%	12.93%	19.91%	23.84%	26.46%
26.46%									

Figure F-3. Cash Flows for Return on Equity

RETURN ON INVESTMENT									
Construction Year	1								
Operating Year		1	2	3	4	5	6	7	8
Book Value	$290,000	$207,143	$147,959	$105,685	$64,257	$22,828	$0	$0	$0
Depreciation (by straight-line)		$41,429	$41,429	$41,429	$41,429	$41,429	$41,429	$0	$0
Depreciation (by double DB)		$82,857	$59,184	$42,274	$30,196	$18,359	$6,522	$0	$0
Depreciation		$82,857	$59,184	$42,274	$41,429	$41,429	$22,828	$0	$0
CASH FLOWS									
Construction Year	1								
Operating Year		1	2	3	4	5	6	7	8
Revenues		$8,400	$8,820	$9,261	$9,724	$10,210	$10,721	$11,257	$11,820
+ Operating Savings		$82,134	$86,240	$90,552	$95,080	$99,834	$104,826	$110,067	$115,570
Net Revenues		$90,534	$95,060	$99,813	$104,804	$110,044	$115,546	$121,324	$127,390
- Depreciation		$82,857	$59,184	$42,274	$41,429	$41,429	$22,828	$0	$0
Taxable Income		$7,677	$35,877	$57,539	$63,375	$68,616	$92,718	$121,324	$127,390
- Income Tax		$2,610	$12,198	$19,563	$21,548	$23,329	$31,524	$41,250	$43,313
Profit after Tax		$5,066	$23,679	$37,976	$41,828	$45,286	$61,194	$80,074	$84,077
+ Depreciation		$82,857	$59,184	$42,274	$41,429	$41,429	$22,828	$0	$0
After-Tax Cash Flow		$87,924	$82,862	$80,250	$83,256	$86,715	$84,022	$80,074	$84,077
Cash Flow for ROI	($295,000)	$87,924	$82,862	$80,250	$83,256	$86,715	$84,022	$80,074	$84,077
Net Present Value	($295,000)	($218,545)	($155,889)	($103,123)	($55,521)	($12,408)	$23,917	$54,019	$81,504
Return on Investment		-70.20%	-30.04%	-7.77%	5.25%	13.20%	17.98%	20.96%	23.08%
23.08%									

Figure F-4. Cash Flows Based on Equity Financing

This Appendix lists reference material that may be helpful to you as you develop your pollution prevention program. The list is divided into the following sections:
- U.S. EPA reports
- state environmental agency reports
- reports by other U.S., regional, and local agencies
- foreign and international agency documents
- industrial and professional societies; universities; corporations reports
- books
- journal articles

The mailing addresses and telephone numbers for the U.S. EPA and the state environmental agencies are listed in Appendix D.

U.S. ENVIRONMENTAL PROTECTION AGENCY

The Pollution Prevention Research Branch maintains a listing of its current projects and publications. Contact the U.S. EPA Risk Reduction Engineering Laboratory, Cincinnati, Ohio.

Achievements in Source Reduction and Recycling for Ten Industries in the United States, EPA/600/2-91/051**

On this and subsequent pages,
*   Available from National Technical Information Service as part of a five-volume set, NTIS No. PB-87-114-328. (703) 487-4650
**   Available from U.S. EPA CERI Publications Unit, 26 West Martin Luther King Drive, Cincinnati, OH 45268. (513) 569-7562.

Electronic Information Exchange System (EIES) — User Guide, Version 1.1, U.S. EPA Pollution Prevention Information Clearing House (PPIC), EPA/600/9-89/086

The Environmental Challenges of the 1990's, Proceedings of the International Conference on Pollution Prevention: Clean Technologies and Clean Products, EPA/600/9-90/039.

Industrial Pollution Prevention Opportunities for the 1990s, EPA/600/891/052**

Pollution Prevention Benefits Manual, Phase II. October, 1989. Draft available from U.S. EPA Pollution Prevention Information Clearing House (PPIC).

Pollution Prevention 1991: Progress on Reducing Industrial Pollutants, EPA-21P-3003.

Powder Coatings Technology Update, EPA-450/3-89-33.

Total Cost Assessment: Accelerating Industrial Pollution Prevention Through Innovative Project Financial Analysis, with Applications to the Pulp and Paper Industry, Report Prepared by the Tellus Institute, December 1991.

A series of reports on waste minimization:
- *Waste Minimization: Environmental Quality with Economic Benefits*, 2nd. ed., April 1990, EPA/530/SW-90/044.
- *Waste Minimization — Issues and Options*, Vols. I-III EPA/530/SW-86/041 through /043 (Washington, D.C.: U.S. EPA, 1986)*
- *Report to Congress: Waste Minimization*, Vols. I and II. EPA/530/SW-86/033 and /034 (Washington, D.C.: U.S. EPA, 1986)**

A series of manuals** describe waste minimization options for specific industries. This is a continuing series which currently includes the following titles:

- *Guide to Pollution Prevention: The Pesticide Formulating Industry*, EPA/625/7-90/004.
- *Guide to Pollution Prevention: The Paint Manufacturing Industry*, EPA/625/7-90/005.
- *Guide to Pollution Prevention: The Fabricated Metal Products Industry*, EPA/625/7-90/006.
- *Guide to Pollution Prevention: The Printed Circuit Board Manufacturing Industry*, EPA/625/7-90/007.
- *Guide to Pollution Prevention: The Commercial Printing Industry*, EPA/625/7-90/008.
- *Guide to Pollution Prevention: Selected Hospital Waste Streams*, EPA/625/7-90/009.
- *Guide to Pollution Prevention: Research and Educational Institutions*, EPA/625/7-90/010.
- *Guide to Pollution Prevention: The Photoprocessing Industry*, EPA/625/7-90/012.
- *Guide to Pollution Prevention: The Automotive Repair Industry*, EPA/625/7-91/013.
- *Guide to Pollution Prevention: The Fiberglass Reinforced and Composite Plastics Industry*, EPA/625/7-91/014.
- *Guide to Pollution Prevention: The Marine Maintenance and Repair Industry*, EPA/625/7-91/015.
- *Guide to Pollution Prevention: The Automotive Refinishing Industry*, EPA/625/791/016.
- *Guide to Pollution Prevention: The Pharmaceutical Industry*, EPA/625/7-91/017.

STATE ENVIRONMENTAL PROTECTION AGENCIES

Alaska Health Project

Wigglesworth, D. *Profiting from Waste Reduction in Your Small Business.* 1988, 46 pp.

On-site Consultation Audit Reports for facilities of the following types:
- aviation facility
- dairy foods processor
- dry cleaner
- fur dressing and tanning shop
- high school
- laboratory facility
- oil field service company
- photofinishing shop
- plastic bottle making/chemical manufacturing
- regional hospital
- seafood processing plant
- secondary seafood processor

Waste Reduction Tips for:
- all businesses
- dry cleaners
- local governments
- newspaper manufacturers
- photofinishers
- print shops
- vehicle repair shops

California Environmental Protection Agency

Alternative Technologies for the Minimization of Hazardous Waste, July 1990.

Alternative Technology for Recycling and Treatment of Hazardous Waste: 3rd Biennial Report, 1986.

Economic Implications of Waste Reduction, Recycling, Treatment and Disposal of Hazardous Wastes: Fourth Biennial Report, July 1988, 126 pp.

Guide to Solvent Waste Reduction Alternatives, October 1986.

Waste Minimization for Hazardous Materials Inspectors: Module I, Introductory Text with Self-Testing Exercises, January 1991, 114 pp.

Waste Minimization Assessment Procedures: Module II.
Unit 1: Waste Minimization Assessment Procedures for the Inspectors with Self-Testing Exercises.
Unit 2: Waste Minimization Assessment Procedures for the Generator

Waste Minimization for the Metal Finishing Industry: Module III.

Waste Minimization: Small Quantity Generators at Los Angeles International Airport. February, 1991.

Various industry-specific checklists; representative titles include:
Printed Circuit Board Manufacturers, February 1991.
Waste Reduction for the Pesticide Formulating Industry, March 1989.
Waste Reduction for the Aerospace Industry, April 1990.
Waste Minimization for Metal Finishers, February 1991.
Waste Minimization for Automotive Repair Shops, February 1991.
Waste Reduction for the Commercial Printing Industry, August 1989.
Waste Reduction Can Work for You!, April 1990.
Waste Reduction for Paint Formulators, December 1989.

Connecticut Technical Assistance Program

Waste Minimization and Pollution Prevention: Self-Audit Manual — Metal Finishing, prepared by Integrated Technologies, Inc., September 1990.

Minnesota Technical Assistance Program

Final Report on the Internship served at Gage Tool Company, 1985.

Minnesota Waste Reduction Institute for Training and Applications Research, Inc. (WRITAR)

Minnesota Guide to Pollution Prevention Planning, February 1991.

Survey of State Legislation. March, 1992.

Survey and Summaries of State Legislation Relating to Pollution Prevention, January, 1991.

North Carolina Department of Environment, Health, and Natural Resources

General and Program Information:
- *Case Summaries of Waste Reduction by Industries in the Southeast*
- *Developing and Implementing a Waste Reduction Program*
- *Pollution Prevention Challenge Grant Information*
- *Waste Reduction Techniques: An Overview*
- *Handbook for Using a Waste Approach to Meet Aquatic Toxicity Limits*
- *Hazardous Materials in North Carolina: A Guide for Decisionmakers in Local Government*
- *Directory of Industrial and Commercial Recyclers Serving North Carolina Businesses and Communities*
- *Directory of State and Local Contacts for Recycling Information and Assistance*
- List of available audiovisual materials

Industry-Specific Information:
- *Water Conservation for Electroplaters: Rinse Tank Design*
- *Water Conservation for Electroplaters: Rinse Water Reuse*
- *Water Conservation for Electroplaters: Counter-Current Rinse*
- *Drag-out Management for Electroplaters*
- *Atmospheric Evaporative Recovery Applied to a Nickel Plating Operation*
- *A Workbook for Pollution Prevention by Source Reduction in Textile*
- *Wet Processing*
- *Identification and Reduction of Pollution Sources in Textile Wet Processing*
- *Identification and Reduction of Toxic Pollutants in Textile Mill Effluents*
- *Water Conservation for Textile Mills*
- *Dye Bath and Bleach Bath Reconstitution for Textile Mills*
- *Ultraviolet Light Disinfection of Water in a Textile Air Washer*
- *Water and Chemical Reduction for Cooling Towers*
- *Small Solvent Recovery Systems*
- *Solvent Loss Control - Things You Can Do Now*
- *Managing and Recycling Solvents*

- *Managing and Recycling Solvents in the Furniture Industry*
- *Waste Reduction Options for Radiator Service Firms*
- *Waste Reduction Options for Automobile Salvage Yards*
- *Garage Owners: Handling of Hazardous and Solid Waste*
- *Pollution Prevention Techniques for the Wood Preserving Industry*
- *Silver Recovery Systems and Waste Reduction in Photoprocessing*
- *Recovery of Volatile Organic Compounds from Small Industrial Sources*
- *Companion Document for the Conference on Waste Reduction for Industrial Air Toxic Emissions*
- *Pollution Reduction Strategies in the Fiberglass Boatbuilding and Open-Mold Plastics Industries*
- *Marine Maintenance and Repair: Waste Reduction and Safety Manual*
- List of available pollution prevention publications for the food processing industry
- *Ten Fact Sheets on Pesticides and Water Quality*
- *Pesticide Rinsate Recycling Facilities Design Guide*
- *Reduction in Pollution from Irrigated Farming*
- *Waste Management Strategies for Hospitals and Clinical Laboratories*
- *Reduction Techniques for Laboratory Chemical Wastes*
- *Reduction of Hazardous Waste from High School Chemistry Labs*
- *Pollution Prevention Pays Instruction Manual for Technical Colleges*

Ohio EPA

Facility Pollution Prevention Planning: A Matrix of the Provisions of Twelve State Laws, October 1990, 25pp.

Oregon Department of Environmental Quality

Benefitting for Toxic Substance and Hazardous Waste Reduction, October 1990.

Tennessee Waste Reduction Assistance Program

Waste Reduction Assessment and Technology Transfer (WRATT) Training Manual, 2nd ed., 1989, 200+ pp.

Writing a Waste Reduction Plan: Charting Your Company's Course Towards Better Waste Management, A How-To Book for Tennessee Generators

OTHER U.S., REGIONAL, AND LOCAL AGENCIES

City of San Jose, CA

Brown, S., R. Kessler, and G. Lynch. *Hazardous Waste Management and Reduction: A Guide for Small- and Medium-Sized Businesses*, 1989.

Great Lakes Rural Network

Maher, J., P. Rafferty, and O. Burch. *The Small Business Guide to Hazardous Materials Management*, 1988, 195 pp.

Local Government Commission

Low Cost Ways to Promote Hazardous Waste Minimization: A Resource Guide for Local Governments, October 1988, 54 pp.

Minimizing Hazardous Wastes: Regulatory Options for Local Governments, December 1988, 31 pp.

Reducing Industrial and Commercial Toxic Air Emissions by Minimizing Waste: The Role of Air Districts, November 1990, 33 pp.

Reducing Industrial Toxic Waste and Discharges: The Role of POTW's, December 1988, 33 pp.

Ohio Department of Natural Resources

Recycling Basics: A Positive Waste Management Alternative for Ohio, 1989, 43 pp.

Southern States Energy Board

Waste Minimization: Workshop Guidance and Sourcebook, July 1990.

U.S. Congress, Office of Technology Assessment

Serious Reduction of Hazardous Waste, 1986.

U.S. Department of Defense

Proceedings of the 1991 DOD/Industry Advanced Coatings Removal Conference, San Diego.

Proceedings of the 1990 DOD/Industry Advanced Coatings Removal Conference, Atlanta.

U.S. Department of Energy

Architect's and Engineer's Guide to Energy Conservation in Existing Buildings. DOE/RL/-01830P-H4.
Volume 1: Energy Use Assessment and Simulation Methods.
Volume 2: Energy Conservation Opportunities.

First Annual International Workshop on Solvent Substitution, Phoenix, December, 1990. (With U.S. Air Force)

Model Waste Minimization and Pollution Prevention Awareness Plan, February 1991, 32 pp.

FOREIGN AND INTERNATIONAL AGENCIES

Dutch Ministry of Economic Affairs. DDU/DOP, Rooseveltstraat 52-56, 2321 BM Leiden, The Netherlands, tel. +3171352500

Manual for the Prevention of Waste and Emissions, Part I, June 1991.

World Bank

The Safe Disposal of Hazardous Wastes, Technical Paper Number 93.

INDUSTRIAL AND PROFESSIONAL ASSOCIATIONS; UNIVERSITIES; CORPORATIONS

Air Pollution Control Association

Cole, G. E. *VOC emission reduction and other benefits achieved by major powder coating operations*. Paper No. 84-38.1, June 25, 1984.

American Society for Testing and Materials

Handbook of Vapor Degreasing. Special Technical Publication 310-A. April, 1976.

Center for Hazardous Materials Research. University of Pittsburgh Applied Research Center, 320 William Pitt Way, Pittsburgh, PA 15238

Hazardous Waste Minimization Manual for Small Quantity Generators in Pennsylvania. April 1987.

Chemical Manufacturers Association. 2501 M Street, N.W., Washington, DC 20037, (202) 887-1100

Improving Performance in the Chemical Industry, Ten Steps for Pollution Prevention, September 1990.

Waste Minimization Resource Manual, 1989.

CH2M Hill. Washington, D.C.

Higgins, T. E. *Industrial Process Modifications to Reduce Generation of Hazardous Waste at DOD Facilities: Phase 1 Report*, 1985.

Dow Chemical. Midland, MI 48674.

Environmental Protection Guidelines for Operations. 18 pp.

Environment Reporter.

Blueprint for National Pollution Prevention Strategy, 56 FR 7849, February 26, 1991.

Hazardous Materials Control Research Institute. Atlanta, GA.

Fromm, C. H. and M. S. Callahan. "Waste Reduction Audit Procedure." *Conference of the Hazardous Materials Control Research Institute*, pp. 427-435, 1986.

HAZTECH International

Fromm, C., S. Budaraju, and S. A. Cordery. "Minimization of Process Equipment Cleaning Waste." *Conference Proceedings of HAZTECH International*, pp. 291-307, Denver, August 13-15, 1986.

3M Corporation. St. Paul, MN.

Ideas — A Compendium of 3M Success Stories

Rutgers University

D. Sarokin. "Reducing Hazardous Wastes at the Source: Case Studies of Organic Chemical Plants in New Jersey." paper presented at *Source Reduction of Hazardous Waste Conference*, August 22, 1985.

Pollution Probe Foundation, Toronto, Ontario.

Campbell, M. E., and W. M. Glenn. *Profit from Pollution Prevention*, 1982.

Pacific Basin Consortium for Hazardous Waste Research, East/West Center, Honolulu, HI.

Waste Minimization: Training Course. November 1990.

BOOKS

Durney, L. J., editor. *Electroplating Engineering Handbook.* 4th ed., New York: Van Nostrand Reinhold. 1984.

Freeman, H. *Hazardous Waste Minimization.* New York: McGraw-Hill, 1990, 343 pp. ISBN 0-07-022043-3.

Glasstone, S. *Energy Deskbook.* New York: Van Nostrand Reinhold, 1983, 453 pp.

Hu, S. D. *Handbook of Industrial Energy Conservation.* New York: Van Nostrand, 1983, 520 pp.

Industrial Waste Audit and Reduction Manual. 2nd ed. Ontario: Ontario Waste Management Corporation, July 1989, 91 pp. ISBN-7729-5851-3.

Van Weenan, J. C. *Waste Prevention: Theory and Practice.* The Hague: CIP-Gegevans Koninklije Bibliotheek, 1990.

JOURNAL ARTICLES

Baumer, A. R. "Making Environmental Audits," *Chemical Engineering* 89(22) 1982, p. 101.

"Cryogenic Paint Stripping." *Product Finishing*, December 1982, pp. 54-57.

Danneman, J. "UV Process Provides Rapid Cure for Compliant Wood Finishes." *Modern Paint and Coatings* 78(2) 1988, pp. 28-29.

Durney, J. "How to Improve Your Paint Stripping." *Product Finishing,* December 1982, pp.52-53.

Fischback, B. C. "Waste Reduction Methodology and Case Histories at Dow Chemical's Pittsburg, California Plant Site." *Environmental Progress* 10(1) 1991, pp. F12-F13.

Geltenan, E. "Keeping Chemical Records on Track." *Chemical Business* 6(11) 1984, p. 47.

Hickman, W. E. and W. D. Moore. "Managing the Maintenance Dollar," *Chemical Engineering* 93(7) 1986, p. 68.

Ingleston, R. "Powder Coatings: Current Trends, Future Developments." *Product Finishing*, August 1991, pp. 6-7.

Kletz, T. A. "Minimize Your Product Spillage." *Hydrocarbon Processing* 61(3) 1982, p. 207.

Krishnaswamy, R. and N. H. Parker. "Corrective Maintenance and Performance Optimization," *Chemical Engineering* 91(7) 1984, p. 93.

Lenckus, D. "Increasing Productivity." *Wood and Wood Products* 87(4) 1982, pp. 44-66, May 1982.

Manik, R. and L. A. Dillard. "Toxics Use Reduction in Massachusetts: The Blackstone Project." *Journal of the Air Waste Management Association* 40(10) 1990, pp. 1368-1371.

"Measuring Pollution Prevention Progress." *Pollution Prevention Review,* Spring 1991, pp. 119-130.

Nelson, K. E. "Use These Ideas to Cut Waste." *Hydrocarbon Processing*, March 1990, pp. 93-98.

Pilcher, P. "Chemical Coatings in the Eighties: Trials, Tribulations, and Triumphs." *Modern Paint and Coatings* 78(6) 1988, pp. 34-36.

Pojasek, R. "Contrasting Approaches to Pollution Prevention Auditing." *Pollution Prevention Review*, Summer 1991, pp. 225-235.

Rimberg, D. "Minimizing Maintenance Makes Money." *Pollution Engineering* 12(3) 1983, p. 46.

Singh, J. B. and R. M. Allen. "Establishing a Preventative Maintenance Program," *Plant Engineering*, February 27, 1986, p. 46.

Smith, C. "Troubleshooting Vapor Degreasers." *Product Finishing,* November 1981, pp. 90-99.

"Waste Minimization for Chlorinated Solvent Users." *ChemAware*, June 1988.

This appendix describes terms specifically related to pollution prevention as they are used in this guide.

Assessment Phase — See Pollution Prevention Assessment Program.

Assessment Team — See Pollution Prevention Assessment Team.

CERCLA — Comprehensive Environmental Response Compensation and Liability Act.

Cross-Media Transfer — Refers to the transfer of hazardous materials and wastes from one environmental medium to another.

Environmental Management Hierarchy — The Pollution Prevention Act of 1990 established a hierarchy as national policy. The hierarchy follows this order: (1) Prevent or reduce pollution at the source wherever feasible. (2) Recycle, in an environmentally acceptable manner, pollution that cannot feasibly be prevented. (3) Treat pollution that cannot feasibly be prevented or recycled. (4) Dispose of, or otherwise release into the environment, pollution only as a last resort.

Feasibility Analysis Phase — The point in a pollution prevention program at which screened waste reduction options are evaluated technically, economically, and environmentally. The results are used to select options to be recommended for implementation.

Implementation Phase — The step in a pollution prevention assessment where procedures, training, and equipment changes are put into action to reduce waste.

Mass Balance — A method of accounting for the quantities of materials produced, consumed, used, or accumulated at; released from; or transported to or from a process or facility as a waste, commercial product or byproduct, or component of a commercial product or byproduct.

Multimedia — Refers to all environmental media (air, land, and water) to which a hazardous substance, pollutant, or contaminant may be discharged, released, or displaced.

Pollution/Pollutants — In this report, the terms "pollution" and "pollutants" refer to all nonproduct outputs, irrespective of any recycling or treatment that may prevent or mitigate releases to the environment.

Pollution Prevention — The use of materials, processes, or practices that reduce or eliminate the creation of pollutants or wastes at the source. It includes practices that reduce the use of hazardous materials, energy, water or other resources, and practices that protect natural resources through conservation or more efficient use.

Pollution Prevention Assessments — Systematic, periodic internal reviews of specific processes and operations designed to identify and provide information about opportunities to reduce the use, production, and generation of toxic and hazardous materials and waste.

Pollution Prevention Assessment Team — A group assembled within a facility to conduct waste reduction assessments. They are selected on the basis of their expertise and knowledge of the process operations.

Pollution Prevention Champion — One or more people designated to facilitate the pollution prevention program by resolving conflicts.

Pollution Prevention Task Force — Overall group responsible for instituting a pollution prevention program, for performing a preliminary assessment, and for guiding the program through the development stages.

Preliminary Assessment/Pre-assessment — A facility survey performed early in the development of a pollution prevention program for the purpose of determining which areas present opportunities for pollution prevention. The information gathered during the pre-assessment is used to prioritize sites for detailed assessment later.

RCRA — Resource Conservation and Recovery Act.

Recycling — Using, reusing, or reclaiming materials/waste, including processes that regenerate a material or recover a usable product from it.

SARA — Superfund Amendments and Reauthorization Act.

Source Reduction — As defined in the Federal Pollution Prevention Act, source reduction is "any practice which 1) reduces the amount of any hazardous substance, pollutant, or contaminant entering any waste stream or otherwise released into the environment (including fugitive emissions) prior to recycling, treatment, and disposal; and 2) reduces the hazards to public health and the environment associated with the release of such substances, pollutants, or contaminants. The term includes equipment or technology modifications, process or procedure modifications, reformulation or redesign of products, substitution of raw materials, and improvements in housekeeping, maintenance, training, or inventory control." Source reduction does not entail any form of waste management (e.g., recycling and treatment). The Act excludes from the definition of source reduction "any practice which alters the physical, chemical, or biological characteristics or volume of a hazardous substance, pollutant, or contaminant through a process or activity which itself is not integral to and necessary for the production of a product or the providing of a service."

Task Force — See Pollution Prevention Task Force.

Toxic Chemical Use Substitution — This term describes replacing toxic chemicals with less harmful chemicals, although relative toxicities may not be fully known. Examples would include substituting a toxic solvent in an industrial process with a chemical with lower toxicity and reformulating a product so as to decrease the use of toxic raw materials of the generation of toxic byproducts.

In this report, this term also includes attempts to reduce or eliminate the use in commerce of chemicals associated with health or environmental risks. Examples include the phaseout of lead in gasoline, the attempt to phase out the use of asbestos, and efforts to eliminate emissions of chlorofluorocarbons and halons. Some of these attempts have involved substitution of less hazardous chemicals for comparable uses, but others involve the elimination of a particular process or product from the market without direct substitution.

Toxics Use Reduction. This term refers to the activities grouped under "source reduction," where the intent is to reduce, avoid, or eliminate the use of toxics in processes and/or products so as to reduce overall risks to the health of workers, consumers, and the environment without shifting risks between workers, consumers, or parts of the environment.

Treatment — Involves end-of-pipe destruction or detoxification of wastes from various separation/concentration processes into harmless or less toxic substances.

Waste — In theory, the term "waste" applies to nonproduct outputs of processes and discarded products, irrespective of the environmental medium affected. In practice, since the passage of RCRA, most uses of the term "waste" refer exclusively to the hazardous and solid wastes regulated under RCRA, and do not include air emissions or water discharges regulated by the Clean Air Act or the Clean Water Act. The Toxics

Release Inventory, TRI, refers to wastes that are hazardous as well as nonhazardous.

Waste Exchange — A central office in which generators who want to recycle valuable components of their waste can register the waste for off-site transfer to others.

Waste Minimization — Source reduction and the following types of recycling: (1) beneficial use/reuse, and (2) reclamation. Waste minimization does not include recycling activities whose uses constitute disposal and burning for energy recovery.

Waste Reduction — This term has been used by the Congressional Office of Technology Assessment and INFORM to mean source reduction. On the other hand, many different groups have used the term to refer to waste minimization. Therefore, care must be employed in determining which of these different concepts is implied when the term "waste reduction" is encountered.

Notes

Notes

Notes

Notes